W0254043

ALLE ZEIT WACH
1842

W. Rudolph (Hrsg.)

Therapie der koronaren Herzerkrankung

Aktuelle Aspekte

Mit 46 Abbildungen und 24 Tabellen

Springer-Verlag Berlin Heidelberg New York
London Paris Tokyo Hongkong

Prof. Dr. med. W. Rudolph
Klinik für Herz- und Kreislauferkrankungen
Deutsches Herzzentrum
Lothstr. 11
D-8000 München 2

CIP-Titelaufnahme der Deutschen Bibliothek

Therapie der koronaren Herzerkrankung : aktuelle Aspekte / W. Rudolph (Hrsg.). – Berlin ; Heidelberg ; New York ; London ; Paris ; Tokyo ; Hongkong : Springer, 1990

ISBN-13: 978-3-540-52032-0 e-ISBN-13: 978-3-642-75318-3
DOI: 10.1007/978-3-642-75318-3

NE: Rudolph, Werner [Hrsg.]

Gesamtherstellung: K. Triltsch GmbH, Würzburg
2119/3335-543210 – Gedruckt auf säurefreiem Papier – Printed on acid-free paper

Vorwort

Die vorliegende Monographie beinhaltet Referate eines Symposiums über aktuelle Aspekte der Therapie der koronaren Herzerkrankung, das vom 13. 1. bis 15. 1. 1989 in Berlin stattfand.

Experten diskutierten über die Bewertung von Risikofaktoren und Perspektiven der primären und sekundären Prävention, die Rolle von Endothel und Thrombozyten und die daraus sich ergebenden therapeutischen Implikationen bei stabiler und instabiler Angina pectoris, über das Problem der asymptomatischen Myokardischämie und ihrer Behandlung, über die Möglichkeiten und Grenzen der Pharmakotherapie, die gegenwärtigen Konzepte zur Behandlung des akuten Myokardinfarktes sowie über den Stellenwert der perkutanen transluminalen Koronarangioplastie und neuerer Entwicklungen wie die Laser-Angioplastie. Diese Monographie bietet damit einen für die ärztliche Tätigkeit wertvollen Überblick.

Den Autoren sei für die sorgfältige Erstellung der Manuskripte und den Damen und Herren des Verlages für ihre Hilfe bei der Vorbereitung und Ausstattung dieses Buches gedankt. Für die allseitige Unterstützung gilt ein besonderer Dank der Firma Mack, die sowohl das Symposium als auch die Publikation ermöglichte und die mit Hilfe dieses Buches die Information einem größeren Interessenkreis zugänglich macht.

München, März 1990 Werner Rudolph

Inhaltsverzeichnis

Autorenverzeichnis

Arnett, D. K., M.S.P.H.
Division of Cardiology, University of South Florida,
College of Medicine, Box 19,
12901 Bruce B. Downs Blvd., Tampa, Florida 33612, USA

Bassenge, E., Prof. Dr. med.
Institut für angewandte Physiologie der Universität
Hermann-Herder-Str. 7, D-7800 Freiburg

Beeder, C., M.A.
Department of Cardiology, Cedars-Sinai Medical Center,
8700 Beverly Blvd., Los Angeles, California 90048, USA

Biamino, G., Prof. Dr. med.
Universitätsklinikum Rudolf Virchow/Wedding,
Röntgendiagnostisches Zentralinstitut, Zentrum für Laserangioplastie,
Augustenburger Platz 1, D-1000 Berlin 45

Dirschinger, J., Dr. med.
Klinik für Herz- und Kreislauferkrankungen
Deutsches Herzzentrum,
Lothstr. 11, D-8000 München 2

Fleck, E., Prof. Dr. med.
Klinik für Innere Medizin – Kardiologie,
Deutsches Herzzentrum Berlin,
Augustenburger Platz 1, D-1000 Berlin 65

Frantz, E., Prof. Dr. med.
Klinik für Innere Medizin – Kardiologie,
Deutsches Herzzentrum Berlin,
Augustenburger Platz 1, D-1000 Berlin 65

Glasser, S. P., M.D.
Division of Cardiology, University of South Florida
College of Medicine, Box 19,
12901 Bruce B. Downs Blvd., Tampa, Florida 33612, USA

Grundfest, W. S., M.D.
Department of Cardiology,
Cedars-Sinai Medical Center,
8700 Beverly Blvd., Los Angeles, California 90048, USA

Heyden, S., M.D., Ph.D.
Department Community and Family Medicine,
Duke University Medical Center,
Durham, North Carolina 27710, USA

Krülls-Münch, J., Priv.-Doz. Dr. med.
Klinik für Innere Medizin – Kardiologie,
Deutsches Herzzentrum Berlin,
Augustenburger Platz 1, D-1000 Berlin 65

Lemmer, B., Prof. Dr. med.
Zentrum der Pharmakologie,
Johann-Wolfgang-Goethe-Universität,
Theodor-Stern-Kai 7, D-6000 Frankfurt am Main

Lichtlen, P. R., Prof. Dr. med.
Medizinische Hochschule Hannover, Abteilung für Kardiologie,
Konstanty-Gutschow-Str. 8, D-3000 Hannover 61

Litvack, F., M.D.
Department of Cardiology, Cedars-Sinai Medical Center,
8700 Beverley Blvd., Los Angeles, California 90048, USA

Mülsch, A., Dr. med.
Institut für angewandte Physiologie,
Hermann-Herder-Str. 7, D-7800 Freiburg

Oswald, H., Dr. med.
Klinik für Innere Medizin – Kardiologie,
Deutsches Herzzentrum Berlin,
Augustenburger Platz 1, D-1000 Berlin 65

Rudolph, W., Prof. Dr. med.
Klinik für Herz- und Kreislauferkrankungen,
Deutsches Herzzentrum,
Lothstr. 11, D-8000 München 2

Rutsch, W., Priv.-Doz. Dr. med.
Freie Universität Berlin,
Universitätsklinikum Rudolph Virchow/Charlottenburg,
Spandauer Damm 130, D-1000 Berlin 19

Schmutzler, H., Prof. Dr. med.
Freie Universität Berlin,
Universitätsklinikum Rudolph Virchow/Charlottenburg,
Spandauer Damm 130, D-1000 Berlin 13

Schneider, K. A., M.D.
Department Community and Family Medicine,
Duke University Medical Center,
Durham, North Carolina 27710, USA

Segalowitz, J., M.D.
Department of Cardiology, Cedars-Sinai Medical Centers
8700 Beverly Bldv., Los Angeles, California 90048, USA

Singer, P., M.D.
Department Community and Family Medicine,
Duke University Medical Center
Durham, North Carolina 27710, USA

Intervention in Risk Factors of Ischemic Heart Disease

S. Heyden, K. A. Schneider, and P. Singer

Introduction

Intervention in risk factors with the goal of preventing coronary heart disease (CHD) cannot be effective, complete and successful without first eliminating hypercholesterolemia. Hypercholesterolemia is the most pivotal and direct cause of atherogenesis in the coronary arteries. The definition of hypercholesterolemia is gradually undergoing changes towards lower levels than previously anticipated. Some 360000 men in the US MRFIT screening program (nonintervention) were followed for 7 years. The results indicate that the incidence of CHD is 30% lower at cholesterol levels of 150 mg/dl than it is at 200 mg/dl, and it doubles at 250 and doubles again at 300 mg/dl. In the absence of mass hypercholesterolemia as in Japan, CHD is a rare cause of disease or death, in spite of continuing high rates of other risk factors such as hypertension, cigarette smoking and diabetes.

Table 1 summarizes this concept in a simplistic overview, dividing the world populations into those with low and high cholesterol levels (arbitrarily chosen at 160 mg/dl), and the occurrence of CHD in the presence of the three other risk factors. These factors contribute heavily to the burden of CHD in the same industrialized countries in which mass hypercholesterolemia prevails. As will be shown, there is an ominous clustering of multiple risk factors:

Hypertensive patients demonstrate twice the prevalence of hypercholesterolemia compared to their normotensive counterparts. Dyslipidemia includ-

Table 1. Influence of three risk factors on CHD in the absence or presence of elevated cholesterol

Smoking	Hypertension	Diabetes
Mean cholesterol levels <160 mg/dl		
South and Central America:	Afro-Asian Countries:	Japan:
No CHD	No CHD	No CHD
Mean cholesterol levels >160 mg/dl in western countries		
CHD 5× more frequent than in non- or exsmokers	CHD 7× more frequent than CVA in untreated patients	CHD confirmed in autopsies in 65% of patients

CHD, coronary heart disease; CVA, cerebrovascular accident

ing low HDL-C levels and high LDL-C levels are overrepresented in hypertensives. A two- to threefold increase in glucose intolerance is part of the hypertensive disease, to a certain extent explained by obesity, the common cause of this association in many patients.

Cigarette smokers have been studied in great detail, revealing poor physical condition and quite different eating preferences from nonsmokers, contributing to but not explaining lower HDL-C and higher LDL-C values. Smokers on the basis of chronically higher carboxyhemoglobin (CO Hb) levels have elevated hemoglobin and erythrocyte concentrations. Blood coagulates more easily, fibrinogen levels are higher and platelet survival is shortened in smokers compared to non-smokers, leading to a substantially higher tendency to thrombogenesis. In susceptible individuals, smoking induces arrhythmias. In hypertensive smokers, the effectiveness of antihypertensive medication is considerably reduced, as was shown in the MRC and IPPPSH trials (see Tables 4 and 5).

Diabetes emerges as a multifactorial disease which deserves the broadest diagnostic and therapeutic approach, from dyslipidemias to elevated blood pressure levels and obesity, to prevent cardiovascular sequelae.

It is now recognized that a large proportion of affluent populations in the industrialized nations of Europe and North America are affected by combined increased risk factors, and these are intimately related to a mass disturbance of lipid metabolism. Therefore, our priority as physicians must aim at normalizing low HDL-C and high LDL-C levels. The present paper will examine not what is already well known about total cholesterol in its causal relationship to CHD, but issues surrounding the intriguing cardioprotective HDL-C. In the future, high risk persons will be more precisely identifiable by the measurement of both total cholesterol (TC) and HDL-C. Persons found with high TC but normal HDL-C values may avoid needless alarm, though in the past persons with TC below 200 mg/dl might have been falsely reassured without knowing the risk of an HDL-C level below 35 mg/dl.

HDL – Six Questions and Answers

A series of six questions have been raised and will be answered on the basis of our knowledge in the late 1980s. Undoubtedly, more answers will be forthcoming, but an impressive array of new insights allows us to formulate recommendations to improve the chances for correction of abnormal lipid profiles.

1. What Behavioral Factors (Life Style) Influence HDL Levels?

The relationship between HDL-C and achieved educational level was examined in white women and men aged 20–39 years in nine North American populations surveyed by the Lipid Research Clinics Program. Mean HDL-C values were positively associated with reported educational achievement. Among women and men aged 20–39 years, a gradient of HDL-C levels from

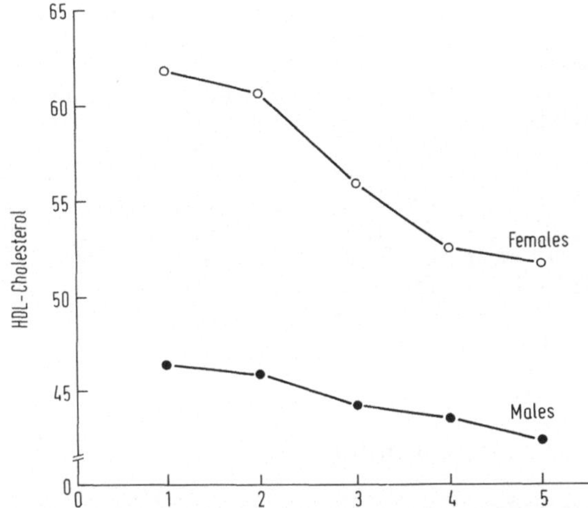

Fig. 1. Mean values of HDL-C by education and sex in participants of the Lipid Research Clinics study aged 20–39 years of age. *1*, Graduate and professional training (college graduate only); *2*, college graduate; *3*, partial college training (at least 1 year); *4*, high school graduate; *5*, partial high school (completed 10th grade). (Tyroler 1980; adapted from Heiss et al. 1980)

lowest to highest educational stratum was found (Fig. 1). The positive association between educational and HDL-C levels was statistically significant in both sexes ($p<0.0002$). "Differences in mean HDL-C between educational strata were in part explained statistically by education-related differences in body mass, alcohol consumption, cigarette smoking, age and, in women, gonadal hormone use. These findings indicate that socioeconomic strata and life style, as measured by level of educational achievement, are associated with HDL-C concentrations in adult populations" (Heiss et al. 1980).

Diet and Alcohol

Ernst et al. (1980) examined the associations of diet and alcohol consumption with levels of HDL-C using cross-sectional data from 4855 white participants, aged 20 years and older, seen at visit 2 of the LRC Prevalence Study. There was a strong positive gradient of HDL levels with reported amount of alcohol intake in both men and women, regardless of type of alcoholic beverage consumed (Table 2).

Weaker inverse relations ($p<0.05$) were found between HDL-C levels and intakes of carbohydrate, sucrose and starch (percent calories and grams). The quality of fat was unrelated to HDL-C levels; similarly, no association was seen between HDL-C and calories, total fat and cholesterol intake (Ernst et al. 1980).

Table 2. Mean HDL cholesterol in relation to alcohol consumption for white men aged 50–69 (adapted in part from Ernst et al. 1980)

Alcohol (ml/day)	Prevalence		Albany*		Framingham*		Honolulu*		San Francisco*	
	n	mean	*n*	mean	*n*	mean	*n*	mean	*n*	mean
0	200	41.87	201	46.28	111	41.40	849	42.18	133	44.38
0.01–16.89	265	47.60	372	47.35	112	44.78	320	44.82	90	45.75
16.90–42.24	144	50.65	260	53.25	111	47.43	354	48.28	40	51.72
42.25–84.50	55	55.25	80	54.57	44	50.14	166	52.20	12	57.75

* As reported by Castelli et al. (1977)

Cigarette Smoking

The relationship between cigarette smoking and HDL-C was examined in 2663 men and 2553 women aged 20–69 years in ten North American populations by Criqui et al. (1980). Male and female smokers had significantly ($p<0.01$) lower HDL levels than nonsmokers, and heavy smokers had lower HDL-C levels than light smokers. Using multiple linear regression analysis to adjust for differences in age, obesity, alcohol consumption and regular exercise increased the differences in HDL levels between smokers and nonsmokers. For men who smoked 20 or more cigarettes/day, adjusted values averaged 5.3 mg/dl (11%) lower than those of nonsmokers ($p<0.01$). Women who used gonadal hormones were analyzed separately from those who did not. In both groups, women who smoked 20 or more cigarettes/day had lower HDL levels than nonsmokers: 9.4 mg/dl (14%) lower in hormone users and 8.6 mg/dl (14%) lower in nonusers (both $p<0.01$) (Fig. 2). "These findings indicate that cigarette smoking is associated with substantially lower levels of HDL-C. Further, this association appears to be dose-dependent, indicating a possible causal relationship between cigarette smoking and lower HDL-C" (Criqui et al. 1980).

Exsmokers had slightly higher HDL levels than neversmokers. Current smokers in both sexes were leanear than ex- or neversmokers and reported higher weekly alcohol consumption; they were less likely to engage in physical activity. The LRC Program Prevalence Study was the first to show that the association between smoking and low HDL levels is independent of the effects of age, hormone use in women, obesity, alcohol use and regular exercise.

Physical Activity

During the LRC North American Prevalence Study, lipoprotein determinations were performed on 2319 white men and 2067 white women aged 20 years or older randomly selected from population surveys by nine clinics in the US and Canada. Participants who reported some strenuous physical activity had

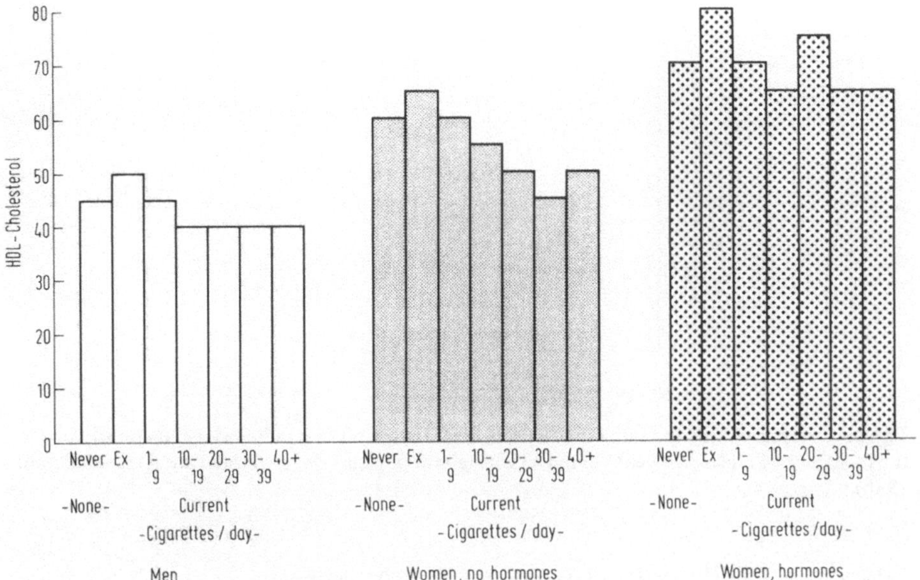

Fig. 2. Mean HDL-C levels plotted against cigarette smoking status according to sex and hormone use in men and women aged 45–69 years of age. Similar but weaker trends were seen in men and women ages 20–44 years, data not shown. (Tyroler 1980; adapted from Criqui et al. 1980)

higher HDL levels than those who reported none, and the more active men aged 30–49 years and active women aged 20–39 had significantly higher values ($p<0.05$). "When HDL-C was adjusted for age, body mass index, alcohol use, cigarette smoking, more active men (47.1 vs. 45.2 mg/dl; $p=0.0001$) and more active women (59.6 vs. 57.7 mg/dl; $p=0.02$) had higher HDL-C levels than their sedentary counterparts. Thus, the association between HDL-C and reported physical activity was, at least in part, independent of other factors that influence HDL-C concentration" (Haskell et al. 1980).

2. Is it True that Low HDL Increases Risk of CHD in the Presence of Normal or Elevated Plasma Total Cholesterol? Is it True that High HDL Decreases Risk of CHD in the Presence of Elevated Plasma Total Cholesterol?

Five prospective epidemiological studies may answer these two questions. In the Framingham Study (Castelli et al. 1986) over a 4 year observation period, healthy men and women were divided into persons with HDL levels below 40 and above 60 mg/dl as well as into four subgroups according to total cholesterol (TC) levels (<200, 200–229, 230–259 and ≥260 mg/dl). For reason of clarity, the intermediate HDL levels were not presented in Fig. 3. The inci-

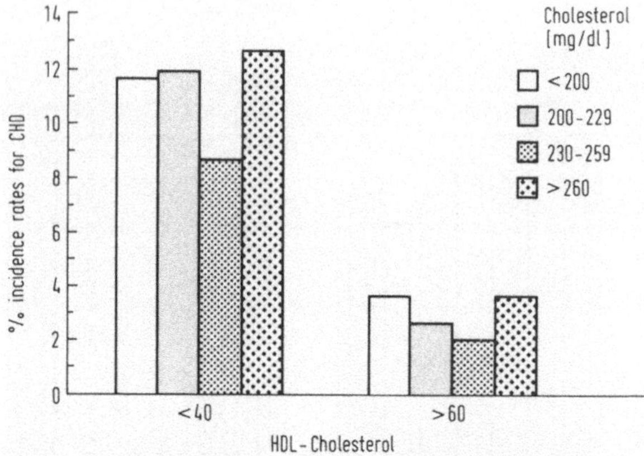

Fig. 3. Comparison of the impact of baseline HDL-C and total cholesterol as predictors of the incidence of CHD in 4 years in the Framingham population. (Adapted from Castelli et al. 1986)

Table 3. Age adjusted CHD mortality rates per 1000 person years in men with previous disease

HDL-C (mg/dl)	MI	GXT	Other
≤34	37.1	16.8	40.1
35–44	26.2	10.5	10.5
≥45	22.2	2.4	5.0

HDL-C, high density lipoprotein cholesterol; MI, myocardial infarction; GXT, positive graded exercise test

dence of CHD is shown to be more influenced by the baseline HDL than by TC levels. The high risk for the development of CHD in a relatively short time period associated with HDL levels below 40 mg/dl is independent of low, slightly, moderately or severely elevated TC levels. Also the cardioprotective effect of high HDL levels appears independent of TC at baseline.

A second Framingham Study concentrated on the over 50 year old group of participants (Abbott et al. 1988), following them over a 12 year time span. Men with TC levels below 192 mg/dl and women below 211 mg/dl with HDL values below 36 (in males) and below 46 mg/dl (in females) experienced a significant increase in myocardial infarction (MI). In a different stratification, in women over 50 years of age with HDL below 46 mg/dl, MI was found six times more often than in women with HDL baseline levels ≥67 mg/dl. In men over 50 years of age 65% of all MI were observed in those individuals with HDL levels below 52 mg/dl.

The third observation is the PROCAM Study, a prospective study of 11094 men and 5000 women, 20–59 years old, conducted in Germany (Assmann and Schulte 1988). Similar to the first Framingham Study, within

4 years, men and women with normal TC levels at baseline and HDL below 35 mg/dl had significantly more CHD than persons with normal HDL levels, *independent* of TC. The 20 year follow-up of the Donolo-Tel Aviv Prospective Coronary Artery Disease Study (Brunner et al. 1987) provided strong evidence for the need to evaluate HDL levels among persons even in the presence of normal TC levels. Men with TC below 200 mg/dl but HDL values below 21 mg/dl (and in women below 23 mg/dl) had a two- to fourfold increased risk for the development of CHD in contrast to men and women with higher HDL levels.

The fifth example is the Lipid Research Clinics Program, a non-intervention, 8½ year follow-up of 2682 men aged 40–69 at baseline (Pekkanen et al. 1988). Age adjusted CHD *mortality rates per 1000* person year in men with previous MI, positive graded exercise test (GXT) or other heart disease manifestations (Other) are shown in Table 3.

Higher mortality rates for men with either previous MI, positive graded exercise test, or other heart disease manifestations were found in those with the lowest HDL when compared with men with higher HDL levels at baseline. TC levels did not show a similarly graded relationship to MI, GXT, or other manifestations of CHD.

3. What Percentage of Individuals with Normal Plasma Total Cholesterol have Low HDL? If These Individuals are at Risk, What is the Magnitude of Risk?

Presently, it is impossible to estimate the number of healthy persons with TC below 200 mg/dl who have low HDL levels. However, it may be assumed from clinical studies of patients with CHD that the percentage is not negligible. Figure 4 describes this special situation in three groups of patients. Ginsburg

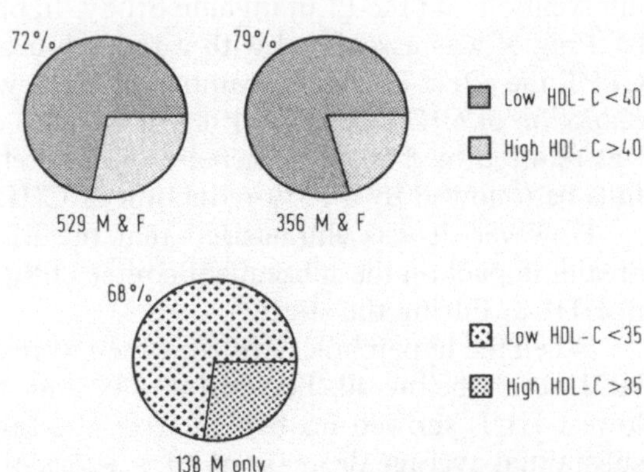

Fig. 4. First estimates of the prevalence of low HDL-C (defined as <40 mg/dl, **a** and **b**; <35 mg/dl, **c**) in CHD patients with total cholesterol less than 200 mg/dl. (Graphs **a** and **b** from Ginsburg et al. 1988; graph **c** from Miller et al. 1988)

et al. (1988) described 529 men and women "with some form of ischemic heart disease" whose TC was below 200 mg/dl; 380 (72%) had HDL below 40 mg/dl. The same group found in 356 men and women with CHD documented by cardiac catheterization and TC below 200 mg/dl, 282 patients (79%) with HDL below 40 mg/dl. Miller et al. (1988) reported on a group of men with CHD, confirmed by coronary angiography; 138 men had TC below 200 mg/dl and 68% of them HDL below 35 mg/dl.

4. What Happens to Plasma HDL Concentration During Drug Therapy, Plasma LDL or Total Plasma Cholesterol? Do Changes in HDL in These Situations Alter Risk of CHD? Which Drugs Should be Studied?

Three placebo controlled randomized drug trials with the purpose of lowering TC and LDL-C levels have specifically addressed the HDL-C problem and consequences for the prevention of CHD and regression of coronary atherosclerosis:

(a) The Lipid Research Clinics Coronary Primary Prevention Trial (LRC-CPPT) was published in 1984 and a more detailed analysis followed in 1986 (Gordon et al. 1986). Under the cholestyramine therapy, HDL levels increased within the first 4 years from 44.3 to 46.8 mg/dl and in the placebo group from 44.4 to 45.6 mg/dl. The investigators concluded that 90% of the achieved reduction of CHD in the active treatment group in comparison to the placebo group is explained by the decrease of LDL-C levels. The contribution of increased HDL-C levels to the reduction of fatal and nonfatal CHD was estimated: with every 1 mg/dl HDL increase from baseline (BL), the incidence of CHD is reduced by 3.5% in the placebo group. This result appears to confirm two other observational studies, the LRC Prevalence Study without intervention and the Framingham Study. In both studies, an increase of HDL by 1 mg/dl was associated with a reduction of CHD by 3.5%. In the LRC-CPPT, the effect of cholestyramine on HDL was enhanced by an even higher reduction of CHD: for each 1 mg/dl increase of HDL, the incidence of CHD was reduced by 5.5%. An increase of, theoretically, 10 mg/dl in HDL would thus be followed by a 55% reduction of CHD.

However, it was emphasized that the BL HDL levels had considerably greater impact on the subsequent course of the trial than the reported changes in HDL-C during the study.

When the hypercholesterolemic men were divided into subgroups with BL HDL levels below 40, between 40–49 and ≥ 50 mg/dl, the group with the lowest HDL showed no benefit from the lipid lowering drug: in spite of a substantial average decrease of LDL levels of 17.4 mg/dl, the coronary morbidity and mortality, both suspected and confirmed, was three times higher in the control group than in the cholestyramine treated group with BL HDL ≥ 50 mg/dl (Fig. 5). Similarly, among the placebo treated patients, the sub-

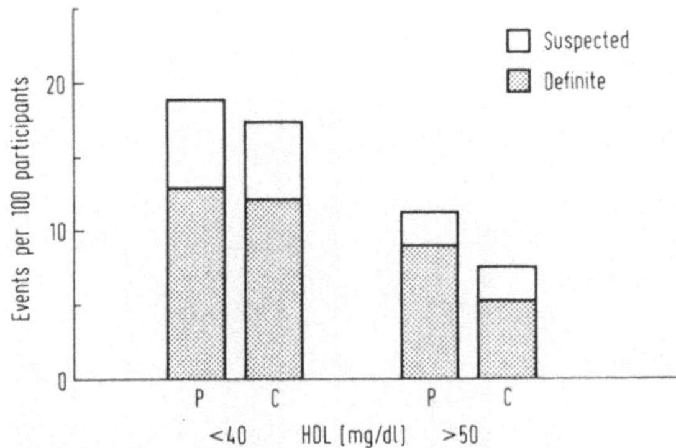

Fig. 5. The 7 year incidence of suspected and/or definite CHD, depending on baseline HDL-C levels in two treatment groups of the Lipid Research Clinics Coronary Primary Prevention Trial. The incidence of CHD in subjects with HDL-C levels between 40 and 49 mg/dl was intermediate, data not shown. *P*, placebo; *C*, cholestyramine. (From Gordon et al. 1986)

group with lowest HDL levels had an almost three times higher incidence of CHD than men with normal HDL levels. Gordon et al. (1986) concluded that the incidence of CHD was 13% higher in the below 40 mg/dl HDL group than could have been expected from the LDL lowering effect of cholestyramine. On the other hand, the incidence of CHD was 25% lower than would have been anticipated under cholestyramine therapy in men whose HDL at BL was ≥50 mg/dl. The predictive power of BL HDL levels for the incidence of CHD has never been demonstrated in a drug trial prior to the LRC-CPPT.

(b) The Helsinki Heart Study described the effect of Gemfibrozil on serum lipids and clinical consequences of CHD (Manninen et al. 1988). The active treatment and placebo groups were divided into two classes according to median HDL value at BL of 39 mg/dl (1.20 mmol/l). The incidence rates of CHD, standardized for differences in smoking and hypertension, were 27 per 1000 men for those with an HDL value above the median and 29 per 1000 for those with an HDL below the median in the Gemfibrozil treated group. Corresponding rates for men in the placebo group were 31 and 50 per 1000. From this difference in incidence rates, the authors concluded that the response to Gemfibrozil was significantly greater among men with low HDL levels at BL than among those with HDL values above the median. Averaging over the 5 years of the trial, Gemfibrozil therapy lowered TC values by 10%, LDL by 11%, and triglycerides by 35% and raised HDL levels by 11% as compared with the placebo. Figure 6 depicts the difference in the impact of Gemfibrozil on the three hyperlipoproteinemia types of Fredrickson. LDL levels were lowered by 11% from BL in Type II A, by 4% in Type II B patients, and did not change in Type IV patients. Responses of HDL were similar in all three Fredrickson types. Reduction of CHD incidence was lowest in Type II A – for which no explanation could be found.

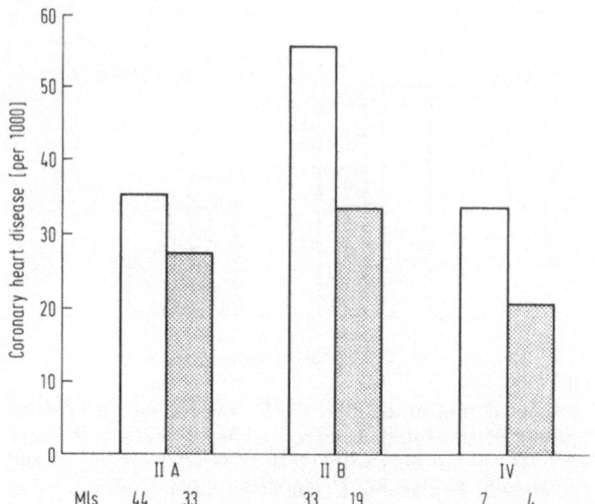

Fig. 6. Differences of impact of treatment over 5 years with Gemfibrozil on hypercholesterolemic men with three Fredrickson types in the Helsinki Heart Study. *Open bars,* placebo; *black bars,* Gemfibrozil; *MI;* myocardial infarction. (From Manninen et al. 1988)

The independent contribution to the reduction of CHD by each lipoprotein response to treatment with Gemfibrozil was estimated: HDL changes were responsible for a 23% reduction, LDL changes alone for a 15% reduction, but both changes occurring simultaneously in HDL and LDL were estimated to effect a 28% reduction of CHD.

(c) The Cholesterol-Lowering Atherosclerosis Study (CLAS, Blankenhorn et al. 1987) followed 80 male patients with MI who had had a coronary bypass operation under combined colestipol-niacin therapy and strict diet and coronary angiography for 2 years. The placebo group ($n = 82$ patients) received only a similarly restricted diet. Weight reduction in the active treatment group amounted to no more than 3 kg and all patients were non- or exsmokers prior to the trial. Therefore, the profound increase in HDL levels by 37% from BL, i.e., from a mean of 45 mg/dl to 61 mg/dl can only be explained by the medication (placebo group showed a slight increase of 2%). TC levels were decreased by 26% from BL and LDL by 43% from BL, i.e., from 171 to 97 mg/dl. Stabilization of an otherwise relentless progressive atherosclerotic process in the coronary arteries was achieved in 30 out of 80 actively treated patients (45%), and regression of lesions was seen in 13 out of 80 patients (16%).

In the placebo group, under the influence of dietary changes no new lesions were found in 78%, but regression was recorded only in 2 out of 82 patients. Blankenhorn et al. (1988) demonstrated that in those 64 of the 82 placebo treated men who did not develop new coronary lesions in 2 years, marked dietary changes had taken place. They reduced the intake of saturated and monounsaturated fats significantly ($p < 0.001$), increased the consumption of polyunsaturated fats ($p < 0.001$), and also reduced the total fat intake from 32% at BL to 28% and dietary cholesterol from 258 to 155 mg/dl. Average cholesterol levels in these 64 men were reduced by 13 mg/dl and LDL by 11 mg/dl. Diet adjustments of these patients reflected voluntary food choices in response to computer assisted diet counseling.

From these three studies it appears that nicotinic acid and fibrates produce an increase of HDL levels and should be studied in monotherapy and combined drug treatment regimens, particularly in patients with low HDL-C.

5. What are the Genetic Components or Biochemical Elements that Result in Low HDL? Which Factors Act Independently of LDL and Plasma Total Cholesterol? Which Animal Models Provide Insight into Genetic and Biochemical Factors?

Genetic Influences

In the Bogalusa Heart Study (Freedman et al. 1985) several racial differences were apparent and were statistically significant ($p < 0.05$). Black children had higher HDL-C levels than white children (61.0 mg/dl vs. 53.6 mg/dl). Although blacks also had higher levels of apo A-1 than whites (141.9 mg/dl vs. 132.3 mg/dl), only for boys was the difference statistically significant: 144.5 mg/dl for black boys vs. 129.4 mg/dl for white boys, but 139.3 mg/dl for black girls vs. 135.4 mg/dl for white girls.

Similar findings were reported from the LRC Program Prevalence Study in 424 blacks and 2429 whites. The causes of the differences in lipoprotein fraction cholesterol distribution in LRC populations have not been determined. The HDL-C comparisons between blacks and whites were adjusted for age, body mass, alcohol consumption, smoking, and differences in education achievement. Under these circumstances, large differences in HDL-C persisted between black and white males.

The data suggest increasing black-white lipoprotein fraction differences with increasing age from childhood to young adulthood (Tyroler et al. 1980). "There may be a delayed expression of a trait under genetic control; indeed, the higher HDL-C levels in blacks, particularly in males, first appear in the preschool years, increase in the school years and become most marked in the later part of the second decade of life . . . We cannot distinguish between genetically and environmentally determined differences in lipoproteins from these observations alone . . . Difference in HDL-C and complementary differences in VLDL-C and LDL-C have been observed in blacks and whites in the US, England, and the Caribbean, in rural and urban locales, and in children and adults. This consistency points to the pervasiveness of the determinants of black-white lipid and lipoprotein differences across these diverse populations and settings and makes a genetic hypothesis plausible" (Tyroler et al. 1980).

Biochemical Influences

In the LRC Program Prevalence Study, the influence of several drugs on HDL levels was examined (Wallace et al. 1980). A 20% reduction in mean HDL-C levels in female propranolol users, though somewhat difficult to interpret, seemed important. Male propranolol users showed a reduction of 7.6 mg/dl, but this was not significant. They found no significant alteration in HDL-C levels associated with thiazide or chlorthalidone use. Despite reports that TC and triglyceride levels are increased in patients taking these agents, the negative findings with regard to HDL-C levels and these diuretics are in keeping with reports from HDFP and MRFIT.

Wallace et al. found a small but significant decrease in HDL-C levels associated with benzodiazepine use in males. Benzodiazepine derivatives are widely prescribed in Western countries and affect lipid metabolism. Under the influence of androgen hormones, some athletes have experienced extremely low HDL levels as would be expected from studies reported in the following section. Fish oil capsules have been shown to reduce triglyceride levels, and, in some studies to increase HDL levels. However, this has not been a consistent finding. Under discussion is the fact that some experiments with fish oil capsules have demonstrated an increase in LDL-C levels, while fish consuming populations have low LDL-C levels and normal or high HDL-C levels (see Table 8).

No animal models are available at the present to test the HDL hypothesis. It has not been possible to reduce HDL levels in animal species and/or to raise them. Some species are particularly vulnerable to cholesterol feeding, especially rabbits, monkeys, and pigs, while others appear totally resistant, e.g., rats (with notoriously high HDL levels), dogs, and cats. Causes for these species differences have not been elucidated.

6. What is the Age-Related Pattern of HDL Concentrations? Do Low HDL Levels in Childhood Correlate with HDL Levels in Adults? Is HDL Level in Childhood a Predictor of CHD Risk?

Some of the answers to these questions may be derived from the LRC Study in children and young adults ($n=1639$ white males and females aged 6–25 years).

The data from the LRC Program Prevalence Study, together with evidence from previous studies, confirm that HDL-C declines markedly in adolescent white males. This decline suggests that sex hormones have a differential effect on HDL-C in adolescent males and females.

Figure 7 shows a sharp decline in HDL-C during puberty among white males, from a mean of 55 mg/dl at age 10 to a mean of 44 mg/dl at age 20, while female experience a small increase, from a mean of 51 mg/dl at age 6 years to a mean of 55 mg/dl at age 25 years. Thereafter no further changes occur with

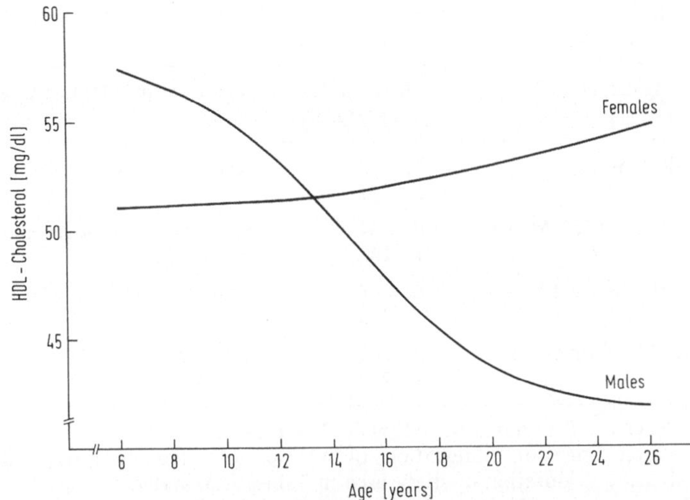

Fig. 7. The relationship between mean HDL-C and age estimated by a nonlinear model as measured in the Lipid Research Clinics Visit 2 Survey examination. (Tyroler 1980; from Beaglehole et al. 1980)

age. While the mean HDL-C levels are significantly higher in males aged 6–10 years than in females, the reverse is true for females 18–25 years old compared to males of the same age ($p<0.001$; Beaglehole et al. 1980).

In females, no sequential changes in either HDL-C or LDL-C were found with sexual maturation (Morrison et al. 1979), suggesting that the male–female difference in HDL-C that appears during maturation may be associated with increased testosterone production in males (Beaglehole et al. 1980).

Data derived from earlier cross-sectional studies in children show that the mean LDL-C/HDL-C ratio in children is much lower than that found in adults, suggesting a transition from childhood to adulthood with respect to lipoprotein interrelationship. "In fact, it has been found in the Bogalusa Heart Study that the LDL-C/HDL-C ratio rapidly increases during puberty in white males and that this is due primarily to a decrease in HDL-C levels. These findings are consistent with the increased risk of coronary artery disease in white males" (Freedman et al. 1985).

Mjos et al. (1977), reporting HDL-C levels in 220 Norwegian children from the Tromso Heart Study, found that males aged 6–9 years had higher levels than females and that they declined significantly with age in males up to age 19 years but not in females. A decrease in HDL-C among boys during the pubertal years was described by Ellefson et al. (1978) in a cross-sectional study of 2421 children aged 6–18 years from Rochester, Minnesota. In males, mean HDL-C declined from 63 mg/dl at 6–10 years of age to 41 mg/dl at 15–18 years of age; HDL did not decline in females.

According to the LRC Study in children and young adults, weight is the variable most consistently associated with HDL-C in both males and females

Table 4. Four hypertension intervention trials in patients with untreated elevated cholesterol levels at baseline and at termination

Study center[a] (*n*)	Mean cholesterol (mg/dl)	Incidence of CHD compared to controls
Oslo (784 M)	B: 275 T: 277	Same as in usual care
IPPPSH (6357 M, F)	B: 234 T: 242	Same in beta-blocker and diuretic treatment groups
HEP (884 M, F)	B: 264 T: 265	Same as in observation group
HAPPHY (6569 M)	B: 242 T: 237	Same in beta-blocker and diuretic treatment groups

B, baseline; T, termination; M, male; F, female

[a] Eleven names or abbreviations of hypertension intervention trials; chief author; (year of publication); duration of study used in Tables 4, 5, and 7.

Oslo	The Oslo Study. Treatment of mild hypertension; Helgeland (1989); 5 years
IPPPSH	International Prospective Primary Prevention Study in Hypertension; Collaborative Group (1985); 3–5 years
HEP	Hypertension in Elderly Patients in Primary Care; Cooper and Warrender (1986); 4 years
HAPPHY	Heart Attack Primary Prevention in Hypertension; Wilhelmsen et al. (1987); 4 years
MRFIT	Multiple Risk Factor Intervention Trial; Research Group (1982); 6 years (CHD mortality)
Australia	Australian National Blood Pressure Study; Management Committee (1984); 1 year (morbidity and mortality)
EWPHE	European Working Party Hypertension in the Elderly Trial; Amery et al. (1985); 8 years
MRC	Medical Research Council Working Party (1985); 5 years (subgroup of MRC trial, 1985)
HDFP	Hypertension Detection and Follow-up Program; Schneider et al. (1987) and HDFP Cooperative Group (1988); 8 years
Gothenburg	Primary Prevention Trial in Gothenburg, Sweden; Samuelsson et al. (1987); 12 years
MAPHY	Metoprolol Atherosclerosis Prevention in Hypertension Study; Wikstrand (1988); 5 years

over the age of 11 years and is the variable most highly correlated with HDL-C in males during the period when HDL-C levels appear to be decreasing (Beaglehole et al. 1980). Eighteen to twenty-five year old females show consistent inverse associations of total carbohydrate and sucrose intake with HDL-C. The inverse relationship between HDL-C and sucrose intake has been found in 10 year old children from Bogalusa (Frank et al. 1978) and in the LRC adult population (Ernst et al. 1980).

In response to the above posed questions, the authors conclude: "The consistent inverse associations between smoking, sucrose intake and HDL-C indicate that the well-documented effects of these variables on HDL-C levels in older adults are already apparent in adolescents and young adults. Thus, the

Table 5. Four hypertension intervention trials in patients stratified according to "low" or "high" cholesterol levels at baseline

Study center[a] (*n*)	Mean cholesterol (mg/dl)		Incidence of CHD in two cholesterol groups (per 1000)
MRFIT (1808 M non-smokers)	<250		3
	>250		19
MRFIT (2211 M smokers)	<250		22
	>250		30
Australia (1721 M, F)	<220		11
	>220		21
EWPHE (416 M, F)	"Low"		"lower"
	"high"		"higher"
MRC (9048 M) non-smokers	Placebo	"low"	6.2
		"high"	9.5
	Diuretics	"low"	5.1
		"high"	11.4
	Beta-blockers	"low"	4.3
		"high"	6.3
Smokers	Placebo	"low"	10.4
		"high"	17.0
	Diuretics	"low"	8.5
		"high"	18.2
	Beta-blockers	"low"	11.4
		"high"	18.4

M, male; F, female
[a] See Table 4 for listing of study centers

effects on HDL-C of smoking cessation and limitation of sucrose intake in young adults must be studied . . . The safest recommendations are to avoid smoking and overeating. These recommendations can be implemented very early in life and are not aimed solely at controlling HDL-C" (Beaglehole et al. 1980).

Hypertension

A review of the world's leading hypertension intervention trials conducted between 1980 and 1988 reveals a profound impact of antihypertensive therapy on the prevention of stroke but a dismal small effect in the prevention of CHD among well controlled hypertensive patients. Tables 4 and 5 elucidate one of the reasons, the high baseline cholesterol levels and the absence of treatment of hypercholesterolemia in hypertensive patients. Even in the Framingham Offspring Study, age adjusted prevalence of hypercholesterolemia was found to be higher in untreated hypertensive than normotensive young adults ($n=1470$ men and 1792 women; Table 6; Costa et al. 1988).

Table 7 reveals a new trend in hypertension intervention trials, either spontaneous cholesterol reduction simultaneously occurring with hypertension treatment, as in the Stepped Care therapy of HDFP; or the minor decrease of cholesterol levels induced in the metoprolol treatment group of the MAPHY trial; or the Gothenburg Study in which all men were equally reduced in their hypertensive BP levels but four subgroups of men with different achieved

Table 6. Age adjusted prevalence of hypercholesterolemia in normotensive and mildly hypertensive young adults (Costa et al. 1988)

Sex	Age	Hypercholesterolemia (>240 mg/dl)	
		Normotensive (DBP <85 mmHg)	Mild Hypertensive (DBP 90–104 mmHg)
M	15–34	5%	15%
	35–44	15%	25%
F	15–34	4%	11%
	35–44	8%	14%

DBP, diastolic blood pressure

Table 7. Three hypertension intervention trials involving antihypertensive treatment and/or cholesterol reduction

Study center[a] (*n*)	Achieved blood pressure and/or cholesterol reduction	Incidence of CHD
HDFP (10940 M, F) in stepped (SC) and referred care (RC) treatment	SC: DBP<90 mmHg in 1 year RC: DBP<90 mmHg within 6 years Cholesterol 235 reduced to 220 mg/dl in both groups	43/1000* 51/1000
Gothenburg (686 M) 4 subgroups according to *achieved* cholesterol levels	Mean BP 152/93 mmHg 205 mg/dl 240 mg/dl 264 mg/dl 302 mg/dl	Number of events 14 16 20 37
MAPHY (1609 M, beta-blockers)	BP under control in both groups: beta-blocker group: Cholesterol 244–237 mg/dl	Cardiovascular mortality 15/1000**
(1625 M, diuretics)	Diuretic group: 243–244 mg/dl	36/1000

M, male; F, female; DBP, diastolic blood pressure
* Significant reduction; ** fatal CHD highly significantly reduced
[a] See Table 4 for listing of study centers

cholesterol levels at the end of the study emerged. In all three trials, whenever both elevated BP and cholesterol levels were reduced, a reduction in the incidence of CHD ensued.

Cigarette Smoking

The long-term damaging effects of cigarette smoke inhalation on the respiratory tract and carcinogenesis in the kidney and bladder are the same in people of every country. However, in spite of nicotine-CO Hb accumulation in the blood (Fig. 8) of smokers around the world, ischemic heart disease does not appear to be a consequence of the cigarette habit among smokers in Central/South America or Afro-Asian countries. A striking example for this point is the contrast in susceptibility to CHD among Puerto Rican smokers compared with smokers in the Framingham Study (Fig. 9). With increasing numbers of cigarettes smoked per day, the risk of CHD increases only among men in Framingham. The major difference between men living in Puerto Rico and men living in Framingham is diet, which in Framingham leads to higher mean cholesterol levels than in Puerto Rico.

While neither nicotine nor CO Hb causes permanent damage to the coronary arteries, nicotine increases heart rate and blood pressure and, therefore, the consumption of oxygen by the heart muscle. CO Hb reduces the oxygen-carrying capacity of the blood. While a person is smoking, coronary perfusion must increase to meet the greater oxygen demand. However, among smokers with higher average serum cholesterol concentrations in Europe and on the North American continent (compared with smokers in Central/South America or Japan), the likelihood of an impaired coronary artery system enhances the difficulty to increase the coronary blood flow. Therefore, the augmented demand for oxygen may not be met, resulting in myocardial ischemia.

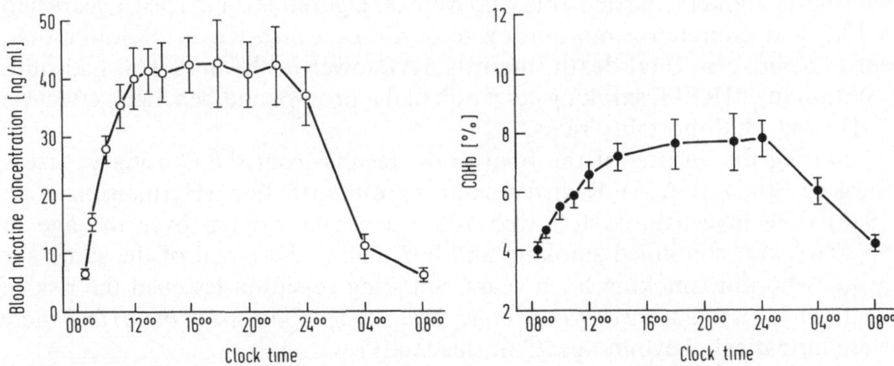

Fig. 8. Persistence of elevated blood nicotine and carboxyhemoglobin (*CoHb*) concentrations (mean ± SEM) in cigarette smokers. Subjects smoked cigarettes every half hour from 8:30 a.m. to 11:00 p.m., for a total of 30 cigarettes per day. (From Benowitz 1988)

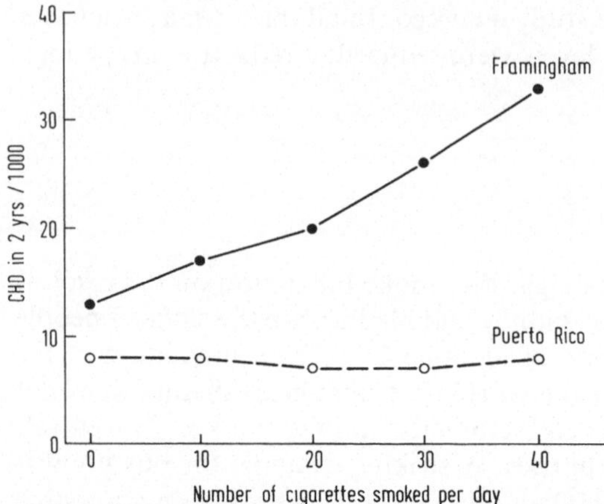

Fig. 9. High risk of CHD in 55 year old male cigarette smokers in Framingham (*closed circles*), average cholesterol level >200 mg/dl, in contrast to low risk of CHD in cigarette smokers in Puerto Rico (*open circles*), average cholesterol level <160 mg/dl. (From Gordon et al. 1974)

The deleterious effect of smoking on HDL-C levels was demonstrated in one of the preceding chapters, as was the profound recovery of HDL-C values among exsmokers. This finding is in agreement with results of numerous studies which have unequivocally demonstrated the reduction of risk of MI in exsmokers. Among healthy exsmokers, the risk reduction to the same low level of MI as in nonsmokers may occur within less than 6 months. On the other hand, the risk reduction may be considerably delayed in patients who quit after having experienced angina pectoris or other conditions such as arrhythmias.

The recent report from MRFIT (Ockene et al. 1988), 10.5 year follow-up of 12866 men aged 35–57 at BL, confirms the notion that cigarette smoking does not cause permanent damage in the coronary arteries. A total of 8194 men (64%) were current smokers, 2799 (22%) were exsmokers and 1873 (14%) were neversmokers. Participants who were off cigarettes for at least 3 years had a 10.5 year cumulative mortality rate of CHD, which was 64% lower than current smokers. Total death rate was 54% lower in the men who had quit. "Within the MRFIT, smoking cessation had a prompt and beneficial effect on CHD and total mortality rates."

During the writing of the final draft, results from the Coronary Artery Surgery Study (CASS) Registry came to our attention (Hermanson et al. 1988). The investigators followed 1086 men and women over the age of 55 years who continued smoking and 807 men and women of the same age group who quit smoking for 6 years. Smoking cessation lessened the risk of death or MI *to the same degree in older as in younger patients with CHD*. (There were no patients beyond age 75 in this study).

The fact that Japanese smoking rates are notoriously higher than those found in the United States sheds an interesting light on the smoking–cholesterol relationship. In several countries and continents where average cholesterol levels are below 160 mg/dl, the traditional risk factors of smoking, hypertension, and diabetes do not contribute to the development of ischemic heart disease (Fig. 9). In contrast, people in countries with high mean serum cholesterol concentrations, above 160 mg/dl, as shown in Table 5 from the MRC in England or from the MRFIT study (1982), are at great risk of developing ischemic heart disease. The same risk factors, smoking, hypertension, and diabetes, pose an excessive risk for ischemic heart disease because of the underlying hypercholesterolemia induced coronary arteriosclerosis. For this reason, the treatment of hypercholesterolemic serum levels assumes a more important role than previously thought, and particularly in smokers, hypertensives, and diabetics.

Diabetes Mellitus

There are five unique features of diabetes mellitus which have emerged from epidemiological research. In fact, epidemiological observations from prospective longitudinal studies such as Framingham and Evans County have probably contributed to a better understanding of the disease and of the potentially most advantageous treatment of the patient.

1. While several U.S. based surveys have demonstrated a formidable reduction of deaths from CHD between the mid-1960s and mid-1980s in all age groups in both sexes, no such decline is seen in the diabetic population. One example: the Allegheny County, Pennsylvania, follow-up study had demonstrated a decline in the CHD death rate by 53% between 1970 and 1984 among 35–44 year old white men, i.e., from 90.6 per 100000 to 42.3 per 100000. On the other hand, an increasing percentage of CHD deaths in the same age group occurred among insulin dependent diabetics, from 6.5% in 1970–1972 to 15.7% in 1984. The author (Kuller 1988) concluded that insulin dependent diabetes is becoming an increasingly important risk factor for CHD death.

2. At all ages, diabetes contributes independently to the major manifestations of atherosclerotic disease, besides CHD, e.g., cerebral infarct, intermittent claudication, and cardiac failure, in both sexes.

3. Diabetic patients tend to carry a larger burden of cardiovascular risk factors than nondiabetic persons. Among these risk factors elevated triglyceride levels, low HDL-C values, high VLDL-C and LDL-C values, and also hypertension, proteinuria, and overweight rank prominently. Much of the excess risk of cardiovascular disease among diabetics can be ascribed to this, but there also appears to be some additional effect that may be attributable to enhanced thrombogenesis. The impact of diabetes on cardiovascular disease is quite variable, depending on the associated cardiovascular risk profile.

4. Diabetes is one of the few metabolic factors of atherogenesis that places women at a higher risk than men. It has a greater impact on all cardiovascular sequelae in women. Both the Framingham and the Evans County Studies have shown severe hypertriglyceridemia in diabetic women (but not in men). The reason for sex differences is unknown.

5. International comparisons of autopsy studies among Japanese diabetic patients and white American and British diabetics reveal an absence of CHD or very low rates of CHD among Japanese diabetics buth high CHD rates in Western societies. While diabetes is just as frequent in Japan as in Western industrialized countries, the traditional Japanese diet may be characterized as representing the ideal noncholesterolemic and nontriglyceridemic diet: calories derived from fat, protein, and carbohydrates comprise, respectively, 25%, 15% and 60% of the diet, in contrast to the present Western diet, i.e., respectively, 40%, 20% and 40%. In addition, the discovery of the beneficial effect of a high fiber diet for both hypercholesterolemia and diabetes will facilitate firm recommendations for the increased use of carbohydrates derived from dietary fibers.

"Since the risk of CHD and cardiovascular disease in the diabetic is greatly influenced by the level of coexistent risk factors, the concept of diabetes control should be broadened to include normalization of the entire cardiovascular risk profile" (Levy and Kannel 1988).

Conclusion

In response to the two questions on intervention of risk factors posed by the organizers of this symposium, "when to start" and "what can be achieved," the answers should be quite clear.

1. The mass hypercholesterolemia of the affluent populations does not begin in middle age, it has its roots in childhood. A unique opportunity to influence large numbers of young people offers itself by presenting issues relating to cholesterol and nutrition in high schools and then testing cholesterol levels with the Reflotron instrument which provides test results instantly. It has been shown that after the age of 16 years (Beaglehole et al. 1980), TC levels increase from year to year and thus, avoiding a potentially deleterious lipid abnormality will lead to the elimination of the most important of all risk factors.

2. As was demonstrated in the discussion of the selected three risk factors, hypertension, smoking, and diabetes, all of them are highly interrelated with dyslipidemias. Without achieving control over this mass phenomenon, no major impact on the prevention of CHD can be expected. There is one exception to this rule: smoking cessation, at least in those individuals who are young, healthy and asymptomatic at the time of giving up the habit, will reduce the chances for a heart attack to the same low risk as experienced by neversmokers. Lowering of cholesterol levels and decreasing the number of smokers in a high risk population will contribute to a greater than 50% reduction of CHD, according to conservative estimates. Additional beneficial effects can be expected by the control of hypertensive blood pressure levels and increased regular physical activity. Avoiding obesity and reducing weight would be helpful in the elimination of obesity induced changes in glucose and lipid metabolism as well as hypertension.

Table 8. Influences of life style and intrinsic factors on HDL levels

Increase in HDL	Decrease in HDL
Life style	
Weight reduction	Obesity, weight gain
Smoking cessation	Cigarette smoking
Physical activity	Physical inactivity
Estrogen	Progesterone, androgens
Alcohol	Anabolic steroids
Fish oil (some studies)	Beta-blockers without ISA
Intrinsic factors	
Low triglyceride levels	Diabetes with increase in TG
Black race	White race
"Longevity" factor	Familial hypercholesterolemia

HDL, high density lipoprotein; TG, triglycerides

This summary is not a proclamation of hope or faith but a description of the dynamic change in CHD mortality over the past 20 years in Australia and the USA. The 40% decline in CHD deaths between 1965 and 1985 involved all age and ethnic groups and both sexes and continues with a 3% fall per year. Thus, the self-destructive life styles in the postwar years with their disastrous consequences of epidemic proportions (Table 8) have made way for life-saving life-style changes in the past 20 years which have already proven to be effective and have set the trend for the future.

References

Abbott RD, Wilson PWF, Kannel WB, Castelli WP (1988) High density lipoprotein cholesterol, total cholesterol screening and myocardial infarction. The Framingham Study. Arteriosclerosis 8:207–211

Amery A, Birkenhager W, Bulpitt C et al. (1982) Influence of antihypertensive therapy on serum cholesterol in elderly hypertensive patients. Acta Cardiol 37:235–244

Assmann G, Schulte H (1988) Ergebnisse und Folgerungen aus der Prospektiven Cardiovaskulären Münster (PROCAM) Studie. In: Assman G (Hrsg) Fettstoffwechselstörungen und koronare Herzkrankheit. MMV Medizin Verlag, München, S 97–131

Beaglehole R, Trost DC, Tamir I, Kwiterovich P, Glueck CJ, Insull W, Christensen B (1980) Plasma high-density lipoprotein cholesterol in children and young adults. The Lipid Research Clinics Program Prevalence Study. Circulation [Suppl IV] 62:83–92

Benowitz NL (1988) Pharmacological aspects of cigarette smoking and nicotine addiction. N Engl J Med 319:1318–1330

Blankenhorn DH, Nessim SA, Johnson RL, Sanmarco ME, Azon SP, Cashin-Hemphill L (1987) Beneficial effects of combined colestipol-niacin therapy on coronary atherosclerosis and coronary venous bypass grafts. JAMA 257:3233–3240

Blankenhorn DH, Johnson RL, El Zein HA, Vailas LI (1988) Dietary fat influences human coronary lesion formation. Circulation [Suppl II] 78:11 (Abstract)

Brunner D, Weisbort J, Meshulam N, Schwartz S, Gross J, Saltz-Rennert H, Altman S, Loebl K (1987) Relation of serum total cholesterol and high density lipoprotein cholesterol percentage to the incidence of definite coronary events: twenty-year follow-up of the Donolo-Tel Aviv prospective coronary artery disease study. Am J Cardiol 59:1271–1276

Castelli WP, Gordon T, Hjortland MC et al. (1977) Alcohol and blood lipids. The Cooperative Lipoprotein Phenotyping Study. Lancet II:152–155

Castelli WP, Garrison RJ, Wilson PWF, Abbott RD, Kalousdian S, Kannel WB (1986) Incidence of coronary heart disease and lipoprotein cholesterol levels. The Framingham Study. JAMA 256:2835–2838

Cooper J, Warrender TS (1986) Randomized trial of treatment of hypertension in elderly patients in primary care. Br Med J 293:1145–1151

Costa ME, D'Agostino RB, Belanger AJ, Kannel WB, Chobanian AV (1988) Mild hypertension and hypercholesterolemia in young people: The Framingham Study. Circulation [Suppl II] 78:567 (Abstract)

Criqui MH, Wallace RB, Heiss G, Mishkel M, Schonfeld G, Jones GTL (1980) Cigarette smoking and plasma high-density lipoprotein cholesterol. The Lipid Research Clinics Program Prevalence Study. Circulation [Suppl IV] 62:70–76

Ellefson RD, Elveback LR, Hodgson PA, Weidman WH (1978) Cholesterol and triglycerides in serum lipoproteins of young persons in Rochester, Minnesota. Mayo Clin Proc 53:307

Ernst N, Fisher M, Smith W, Gordon T, Rifkind BM, Little JA, Mishkel MA, Williams OD (1980) The association of plasma high-density lipoprotein cholesterol with dietary intake and alcohol consumption. Circulation [Suppl IV] 62:41–52

Frank GC, Berenson GS, Webber LS (1978) Dietary studies and the relationship of diet to cardiovascular disease risk factor variables in 10-year-old children. The Bogalusa Heart Study. Am J Clin Nutr 31:328–332

Freedman DS, Srinivasan SR, Voors AW, Webber LS, Berenson GS (1985) High density lipoprotein and coronary artery disease risk factors in children with different lipoprotein profiles: Bogalusa Heart Study. J Chronic Dis 38:327–338

Ginsburg GS, Safran C, Pasternak RC (1988) National Cholesterol Education Program guidelines: the impact of missing HDL. Circulation [Suppl II] 78:382 (Abstract)

Gordon T, Garcia-Palmieri R, Kagan A, Kannel B, Schiffman J (1974) Differences in coronary artery disease in Framingham, Honolulu and Puerto Rico. J Chronic Dis 27:329–344

Gordon DJ, Knoke J, Probstfield JL, Superko R, Tyroler HA (1986) High-density lipoprotein cholesterol and coronary heart disease in hypercholesterolemic men. The Lipid Research Clinics Coronary Primary Prevention Trial. Circulation 74:1217–1225

Haskell WL, Taylor HL, Wood PD, Schrott H, Heiss G (1980) Strenuous physical activity, treadmill exercise test performance and plasma high-density lipoprotein cholesterol. The Lipid Research Clinics Program Prevalence Study. Circulation [Suppl IV] 62:53–61

Heiss G, Haskell W, Mowery R, Criqui MH, Brockway M, Tyroler HA (1980) Plasma high-density lipoprotein cholesterol and socioeconomic status. Circulation [Suppl IV] 62:108–115

Helgeland A (1980) Treatment of mild hypertension: A five year controlled drug trial. The Oslo Study. Am J Med 69:725–732

Hermanson B, Omenn GS, Kronmal RA, Gersh BJ et al. (1988) Beneficial six-year-outcome of smoking cessation in older men and women with coronary artery disease: results from the CASS-registry. N Engl J Med 319:1365–1369

Hypertension Detection and Follow-up Program Cooperative Group (1988) Persistance of reduction in blood pressure and mortality of participants in the Hypertension Detection and Follow-up Program. JAMA 259:2113–2122

IPPPSH Collaborative Group (1985) Cardiovascular risk and risk factors in a randomized trial of treatment based on the beta-blocker oxprenolol: The International Prospective Primary Prevention Study in Hypertension. J Hypertens 3:379–392

Kuller LH (1988) The disappearance of coronary heart disease deaths, aged 35–44; 1970–1986 (Abstract). Circulation 78:89

Levy D, Kannel WB (1988) Cardiovascular risks: New insights from Framingham. Am Heart J 116:266–272

Management Committee of the Australian National Blood Pressure Study (1984) Prognostic factors in the treatment of mild hypertension. Circulation 69:668–676

Manninen V, Elo MO, Frick MH et al. (1988) Lipid alterations and decline in the incidence of coronary heart disease in the Helsinki Heart Study. JAMA 260:641–651

Medical Research Council Working Party (1985) MRC trial on treatment of mild hypertension: principal results. Br Med J 291:97–104

Miller M, Mead L, Kwiterovich PO Jr, Pearson TA (1988) Lipid abnormalities in coronary disease patients with "desirable" cholesterol levels: should we screen all CAD patients for low HDL levels? Circulation [Suppl II] 78:383 (Abstract)

Mjos OD, Thelle DS, Forde OH, Vik-Mo H (1977) Family study of high density lipoprotein cholesterol and the relation to age and sex. The Tromso Heart Study. Acta Med Scand 201:323–329

Morrison JA, Laskarzewski RM, Rauh JL et al. (1979) Lipids, lipoproteins, and sexual maturation during adolescence. The Princeton Maturation Study. Metabolism 28:641–645

Multiple Risk Factor Intervention Trial Research Group (1982) The Multiple Risk Factor Intervention Trial. Risk factor changes and mortality results. JAMA 248:1465–1472

Ockene JK, Kuller LH, Svendsen KH, Meilahn E (1988) The differential effect of smoking cessation on CHD and lung cancer mortality in the multiple Risk Factor Intervention Trial (MRFIT): 10.5 years of follow-up. Circulation [Suppl II] 78:10 (Abstract)

Pekannen J, Linn S, Suchindran CM, Heiss G, Tyroler HA (1988) Total and lipoprotein cholesterol and cardiovascular mortality in men with and without prevalent heart disease. Circulation [Suppl II] 78:281 (Abstract)

Samuelsson O, Wilhelmsen L, Andersson OK, Pennert K, Berglund G (1987) Cardiovascular morbidity in relation to change in blood pressure and serum cholesterol levels in treated hypertension. JAMA 258:1768–1776

Schneider KA, Heyden S, Ford C (1987) Failure to reduce cholesterol as explanation for the limited efficacy of antihypertensive treatment in the reduction of coronary heart disease. Evidence from the Hypertension Detection and Follow-up Program. Nephron [Suppl 1] 47:104–107

Tyroler HA (ed) (1980) Epidemiology of plasma high density lipoprotein cholesterol levels. Lipid Research Clinics Program. AHA Monogr 73

Tyroler HA, Glueck CJ, Christensen B, Kwiterovich PO (1980) Plasma high-density lipoprotein cholesterol comparisons in black and white populations. The Lipid Research Program Prevalence Study. Circulation [Suppl IV] 62:99–107

Wallace RB, Hunninghake DB, Reiland S, Barrett-Connor E, Mackenthun A, Hoover J, Wahl P (1980) Alterations of plasma high-density lipoprotein cholesterol levels associated with consumption of selected medications. The Lipid Research Clinics Program Prevalence Study. Circulation [Suppl IV] 62:77–82

Wikstrand J (1988) Primary prevention in patients with hypertension: Comments on the clinical implications of the MAPHY Study. Am Heart J 116:338–347

Wilhelmsen L, Berglund G, Elmfeldt D et al. (1987) Beta-blockers vs. diuretics in hypertensive men. Main results from the HAPPHY trial. J Hypertens 5:561–572

Vasodilatation durch Nitrate und EDRF

E. Bassenge und A. Mülsch

Einleitung

Die Perfusion des Herzens hängt nicht nur von nervösen und metabolischen Faktoren ab, sondern auch von einer Reihe endothelialer Faktoren, insbesondere auch vom „endothelium derived relaxant factor“ (EDRF), der wahrscheinlich mit dem NO-Radikal identisch ist (Übersicht bei Bassenge u. Busse 1988). Die genauen Freisetzungsmechanismen dieser sog. endothelialen Autakoide, zu denen neben dem EDRF auch das Prostacyclin und der „platelet activating factor“ (PAF) gehören, sind bisher noch nicht genau bekannt. Man weiß nur, daß es eine an die Anwesenheit von Kalzium gebundene, rezeptorgekoppelte Stimulation sowie auch eine mechanisch induzierte Freisetzung von EDRF gibt. Bei intaktem Endothel wirkt eine Reihe von Substanzen dilatierend, und zwar über die Freisetzung von EDRF, der seinerseits zu einer Relaxation der glatten Muskulatur führt. Bei fehlendem Endothel oder auch bei starker Funktionseinschränkung des Endothels, wie z. B. bei Atheromatose, konnte hingegen gezeigt werden, daß diese verschiedenen Stimulatoren durch eine direkte Wirkung auf die glatte Muskulatur zu einer Kontraktion der Gefäßmuskulatur und damit zu einer Widerstandserhöhung führen (Übersicht bei Bassenge u. Busse 1988).

Freisetzung von EDRF durch chemische Reize

Zu den Substanzen, die eine Freisetzung von EDRF bewirken, gehören einerseits neben einer Reihe von vasoaktiven Peptiden verschiedene Blutzellprodukte, wie z. B. Serotonin, ATP, Adenosin, ferner Thrombin und Histamin, die auf spezifische Endothelrezeptoren wirken. Auch im Blut zirkulierende Hormone, wie z. B. Adrenalin, Noradrenalin und Angiotensin II, die zu einer starken Gefäßkonstriktion führen, haben gleichzeitig am Endothel über den EDRF noch einen gewissen dilatierenden Effekt (Übersicht bei Bassenge u. Heusch 1990). Der freigesetzte EDRF bewirkt jedoch nicht nur in der Gefäßwand eine Relaxation und damit eine Gefäßwiderstandserniedrigung; gleichzeitig kommt es zu einer Wirkung im Gefäßlumen auf die Thrombozytenfunktion im Sinne einer Hemmung der Plättchenadhäsion und -aggregation (Busse et al. 1987; Bassenge et al. 1989; Pohl u. Busse 1989). Dabei besteht zwischen dem EDRF und der gleichzeitigen PGI_2-Freisetzung eine potenzierende Wirkung. Bereits die Kombination von unterschwelligen Dosen der beiden Substanzen hat einen signifikanten Effekt auf die Adhäsion und Aggregation der Thrombozyten (s. Abb. 1). Von besonderer Bedeutung ist dabei, daß die Freisetzung dieses Prostaglandins und des EDRF auf verschiedene Stimuli meistens gleichzeitig und im Prinzip sehr ähnlich verläuft.

Abbildung 2 zeigt den zweifachen Einfluß der EDRF-Freisetzung auf die charakteristischen „second messenger": Es kommt zu einer Gefäßdilatation

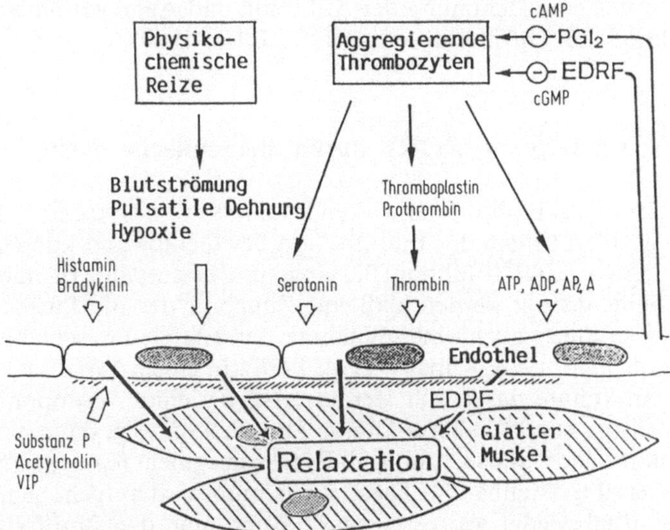

Abb. 1. Bidirektionale Freisetzung von EDRF (NO) aus den Endothelzellen bei Stimulation durch physikalische hydrodynamische Reize (visköse Scherkraft, pulsatile Dehnung), durch aggregierende Plättchen und ihre Freisetzungsprodukte und durch Thrombin. Von abluminal her wirken besonders Substanz P (SP), vasoaktives intestinales Polypeptid (VIP) und Acetylcholin (Ach). Abluminalwärts freigesetzter EDRF (NO) bewirkt Vasodilation, luminalwärts freigesetzter EDRF inhibiert die Plättchenaktivierung, -adhäsion und -aggregation. Gleichzeitig freigesetztes Prostazyklin (PGI_2) potenziert die EDRF-Wirkung

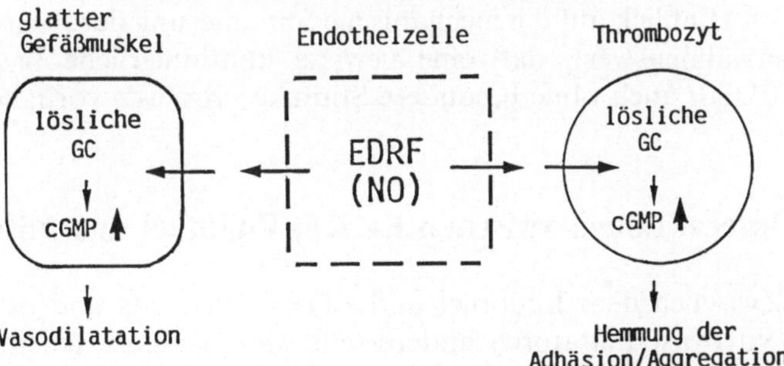

Abb. 2. Die bidirektionale Freisetzung von EDRF (NO) bewirkt einerseits eine Stimulierung der löslichen Guanylatzyklase (sGC) in der glatten Gefäßmuskulatur mit Anstieg von cGMP und daraus resultierender Relaxation, andererseits von sGC in den Thrombozyten ebenfalls mit einem Anstieg von cGMP und daraus folgender Hemmung der Adhäsion und Aggregation

auf der einen Seite und zu einer Plättchenaggregations- und Plättchenadhäsionshemmung auf der anderen Seite. Der sehr gut diffusible EDRF wirkt auf das Schlüsselenzym der Gefäßdilatation, die lösliche Form der Guanylatzyklase, die ihrerseits einen Anstieg von Zykloguanosinmonophosphat (cGMP) in der glatten Muskulatur verursacht; cGMP bewirkt eine Suppression des Kalziumspiegels in der glatten Muskulatur und damit schließlich eine Vasodilatation. Auf der anderen Seite bewirkt der Anstieg des cGMP in den Thrombozyten eine Hemmung der Adhäsion und Aggregation sowie der Releasereaktion der Thrombozyten (Pohl u. Busse 1989).

Freisetzung von EDRF durch mechanische Reize

Die auf das Endothel einwirkende Scherkraft des strömenden Blutes sowie das pulsatile Dehnen des Endothels in der Gefäßwand können zu einer Freisetzung von EDRF führen (Bassenge u. Busse 1988). In einem Bioassaysystem konnte gezeigt werden, daß eine Zunahme des Blutflusses bzw. Pulsatilitätssteigerungen auch eine Zunahme der EDRF-Freisetzung zur Folge haben (Pohl et al. 1986; Rubanyi et al. 1986). In einem Donor-Bioassay-Detektorsystem konnte dargestellt werden, daß bei einer Verdopplung des Blutflusses über das superfundierte Donor-Endothel hinweg, dann eine signifikante Abnahme der Wandspannung im Detektorsegment (einer perfundierten Koronararterie) feststellbar ist. Diese Relaxation wird aufgehoben, wenn der erhöhte Blutfluß wieder gedrosselt wird. Wenn man den Blutfluß konstant läßt, die Pulsatilität jedoch erhöht, kommt es ebenfalls zu einer Relaxation dieses Detektorringes: Durch die pulsatile Dehnung des EDRF-Donor kommt es sofort zu einer Relaxation des Detektorsegments. Gibt man bei diesem Versuch jedoch den EDRF-Inhibitor Hämoglobin dazu, der sich sofort an EDRF bindet und damit EDRF inaktiviert, dann bleibt diese Relaxation aus. Hebt man die Inaktivierung durch Hämoglobin auf (Pohl et al. 1986), kommt es hingegen wieder zu einer Relaxation des Detektorsegments wie zu Beginn des Versuchs. Im Hinblick auf die mechanische Stimulierung der Freisetzung von EDRF ist erwähnenswert, daß eine gewisse kontinuierliche basale Freisetzung von EDRF auch ohne besondere Stimulierungstests vorhanden ist.

Interaktionen zwischen EDRF, Endothel und Nitrovasodilatatoren

Zwischen dem Endothel und EDRF einerseits und dem Endothel und den Nitrovasodilatatoren andererseits gibt es überraschenderweise unterschiedliche Interaktionen (Bassenge u. Stewart 1988), obwohl EDRF bzw. NO eigentlich der aktive Metabolit der Nitrovasodilatatoren ist. In Anwesenheit von einem intakten Endothel sind die Gefäßreaktionen auf Nitroglycerin schwächer ausgeprägt als bei einem Koronarsegment mit denudiertem Endothel. Mit steigenden Nitroglycerindosen sieht man bei einem geschädigtem

Endothel (aber bei noch dehnungsfähiger Gefäßwand) eine stärkere Zunahme des Gefäßdurchmessers, während die dilatative Reaktion auf ansteigende Nitroglycerindosen bei einem Gefäßsegment mit funktionstüchtigem Endothel eher geringer ist (Busse et al. 1989). Diese bei In-vitro-Untersuchungen gewonnenen Erkenntnisse konnten auch unter klinischen Bedingungen nachgewiesen werden. Rafflenbeul et al. (1989) untersuchten bei einem Patientenkollektiv mit mehr oder weniger stark ausgeprägter Koronarsklerose die dilatierende Wirkung von Bradykinin. Nach intrakoronarer Gabe von Bradykinin kam es bei dieser Untersuchung bei einem Teil der untersuchten Gefäße zu einer Dilatation des angiographisch gemessenen koronaren Gefäßdurchmessers durch die endothelvermittelte Gefäßrelaxation. Die Gabe von Nitroglycerin bei diesen Gefäßsegmenten führte jedoch zu einer relativ geringen Gefäßdilatation. Bei einem anderen Teil der Patienten war jedoch die bradykininvermittelte, endotheliale Stimulation der EDRF-Freisetzung gering, so daß es nur zu einer minimalen Gefäßdilatation kam. Bei diesen Patienten bzw. Gefäßsegmenten, bei denen offensichtlich ein funktioneller Defekt des Endothels vorlag, führte die Gabe von Nitroglycerin hingegen zu einer ausgeprägten Gefäßdilatation. Ähnliche Ergebnisse berichteten Drexler et al. (1989) sowie Zeiher et al.(1989), die koronarangiographisch nach Gabe von Acetylcholin die Veränderungen des Gefäßdurchmessers untersuchten. Acetylcholin ist der erste endotheliale Vasodilatator, der 1980 von Furchgott u. Zawadzki beschrieben wurde. Die intrakoronare Gabe von Acetylcholin führte in der Untersuchung von Drexler et al. (1989) bei Patienten mit erheblicher Gefäßerkrankung zu einer starken Konstriktion der Koronargefäße. Die Gabe von Nitroglycerin hingegen führte endothelunabhängig zu einer erheblichen Zunahme des Gefäßdurchmessers. Bei Probanden ohne koronare Gefäßerkrankung bewirkte Acetylcholin hingegen eine endothelabhängige Vasodilatation. Diese Gefäßerweiterung beruht auf einer Dilatation der Arteriolen und damit auf einer Zunahme des Blutflusses (sog. „flußabhängige Dilatation“, Holtz et al. 1983) sowie möglicherweise auch auf einer geringen direkten muskarinergen Wirkung am Endothel (Bassenge u. Busse 1988). Auch bei den gesunden Probanden war erwartungsgemäß eine endothelunabhängige Gefäßerweiterung nach Gabe von Nitroglycerin nachweisbar. Die Gefäßkonstriktion durch Acetylcholin bei Patienten mit geschädigten Koronararterien beruht auf einer direkten Stimulation der unter dem Endothel liegenden glatten Muskulatur, die die indirekte, endothelabhängige Dilatation überspielt. Diese direkte Stimulation ist so ausgeprägt, daß dieser Acetylcholintest bei intrakoronarer Gabe von ansteigenden Dosen offenbar zur Charakterisierung und Differenzierung des Koronarzustandes von Patienten herangezogen werden kann.

Wirkung von EDRF auf die Thrombozyten

Bei In-vitro-Versuchen konnte nachgewiesen werden, daß die Hemmung der Thrombozytenaggregation und -adhäsion durch EDRF dosisabhängig ist.

Mit höchsten Konzentrationen von EDRF kann man die Thrombozytenaggregation vollständig unterdrücken (Busse et al. 1987; Bassenge et al. 1989). Verdünnt man hingegen die EDRF-haltige Lösung so stark, daß praktisch kein EDRF mehr enthalten ist, kann man eine vollständige (z. B. thrombininduzierte) Thrombozytenaggregation beobachten. Auch in tierexperimentellen Untersuchungen am Kaninchenherzen ließen sich diese Effekte nachweisen (Pohl u. Busse 1989). Dabei wurde während normaler Herzarbeit ein Thrombozytenbolus arteriell injiziert, und im venösen Effluat wurden die Plättchen auf ihren cGMP-Gehalt sowie auf ihre Adhäsionsrate hin untersucht. In Abbildung 3 ist auf der Ordinate die cGMP-Konzentration der Thrombozyten dargestellt, sowohl unter basalen Bedingungen wie auch nach intrakoronarer Gabe des EDRF-Stimulators Acetylcholin (Ach). Man kann erkennen, daß die Gabe von Acetylcholin zu einer deutlichen Steigerung der EDRF-Freisetzung führt und damit auch zu einer Verdopplung des cGMP-Gehaltes der Thrombozyten innerhalb einer einzigen Zirkulation, d. h. nach einer Kontaktzeit im Koronarsystem von nur 2–3 s. Wenn man hingegen einen endothelunabhängigen Vasodilatator, wie z. B. Adenosin, injiziert, kommt es nicht zu einer deutlich vermehrten Freisetzung von EDRF und auch nicht zu einer Zunahme des cGMP-Gehaltes der Thrombozyten. Wird andererseits durch die gleichzeitige Gabe von Hämoglobin die Wirkung von EDRF blockiert, ist auch ein Anstieg des cGMP-Gehalts in den Thrombozyten nicht nachweisbar. Die Absenkung des cGMP-Spiegels in den Thrombozyten ist jedoch verbunden mit einer verstärkten Thrombozytenadhäsion und -aggregation. Während unter experimentellen Kontrollbedingungen die Adhäsionsrate etwas mehr als 4% betrug, führte die Ach-induzierte, erhöhte EDRF-Freisetzung zu einer

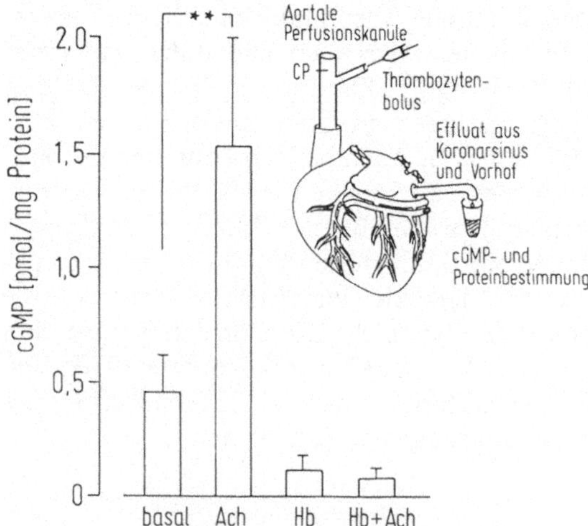

Abb. 3. Stimulierung des Endothels zur verstärkten Freisetzung von EDRF mittels Acetylcholin (Ach) im perfundierten Herzen bewirkt einen Anstieg des cGMP-Gehaltes der Thrombozyten (*Ordinate*) im venösen Effluat. Der EDRF-Inhibitor Hämoglobin (*Hb*) unterdrückt diesen Anstieg und reduziert auch die Wirkung der spontanen, basalen EDRF-Freisetzung mit cGMP-Stimulierung, so daß ein cGMP-Abfall resultiert

Abnahme der Adhäsionsrate der Thrombozyten um mehr als die Hälfte, nämlich auf nur noch 2 %. Mit sehr hohen Acetylcholinkonzentrationen und damit EDRF-Stimulation ließ sich die Adhäsionsrate der Thrombozyten sogar auf weniger als 1 % reduzieren. Ähnliche Effekte sind mit anderen endothelialen Stimulatoren, wie z. B. Bradykinin oder ATP, erzielbar.

Wirksamkeit von EDRF bzw. NO während Nitrattoleranz

Bei längerem Gebrauch von Nitrovasodilatatoren kommt es bekanntlich zu einem Wirkungsverlust dieser Substanzen durch eine Toleranzentwicklung. Wenn man Nitroglycerin in einer Dosis von 1,5 µg/kg min 5 Tage lang parenteral verabreicht, kommt es zunächst zu einer signifikanten Koronardilatation. Bereits am 2. Tag ist jedoch ein starker Wirkungsverlust nachweisbar und nach 3 Tagen werden praktisch nur noch die Kontrollwerte erreicht (Stewart et al. 1987). Wenn man allerdings im Stadium der vollständigen Toleranz gegenüber dem infundierten Nitroglycerin zusätzlich EDRF oder NO freisetzt oder Stimulatoren der NO-Freisetzung anwendet, dann kann der Wirkungsverlust von Nitroglycerin überwunden bzw. ausgeglichen werden. Abbildung 4 zeigt die Ergebnisse einer Untersuchung, bei der periphere Arte-

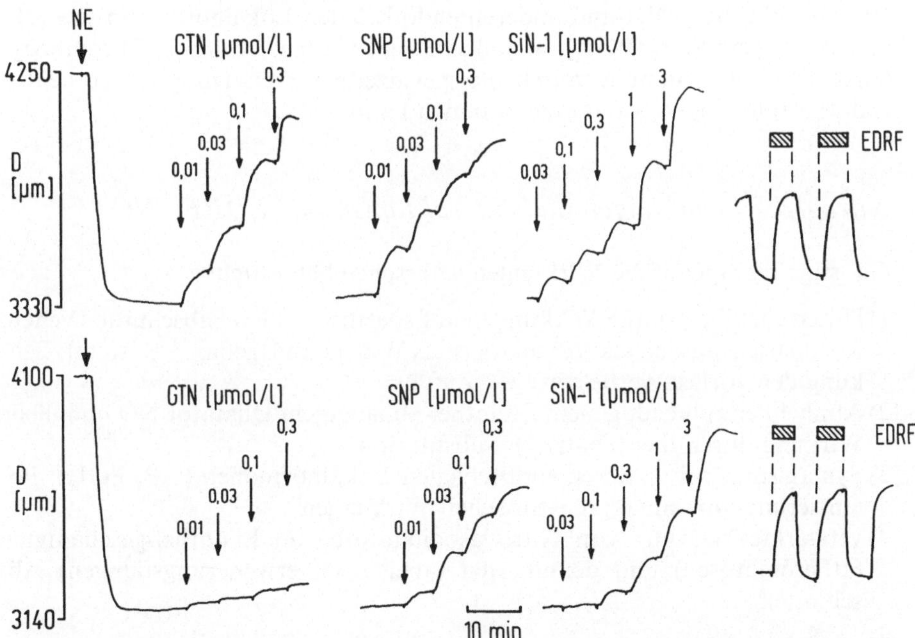

Abb. 4. Dilatation (*D*) von mit Noradrenalin (*NE*; 0,1 µmol/l) vorkontrahierten Arteriensegmenten auf kumulative Konzentrationen von Nitroglycerin (*GTN*), Natriumnitroprussid (*SNP*), Sydnonimin (*SIN-1*) und Perfusion mit EDRF (freigesetzt aus kultivierten Rinderaortenendothelzellen) unter GTN-Toleranz (1 h 0,5 mmol/l GTN, *untere Hälfte*) und Kontroll-Bedingungen (1 h 17 mmol/l Laktose, *obere Hälfte*)

rien und Koronarien unter Kontrollbedingungen und bei vorhandener Toleranz gegen Nitrovasodilatatoren untersucht wurden (Mülsch et al. 1988, 1989a, b). Man kann erkennen, daß unter Kontrollbedingungen mit ansteigenden Dosen von Glyeroltrinitrat, Natriumnitroprussid und SiN-1 sowie bei EDRF-Gabe eine gleichermaßen gute Vasodilatation erreicht werden konnte. In Gefäßen mit Toleranzentwicklung konnte durch die Gabe von EDRF noch eine Gefäßdilatation erreicht werden. Auch die Gabe von Natriumnitroprussid bzw. SiN-1 führte weiterhin zu einer deutlichen Gefäßdilatation. Die Ursache für dieses Phänomen liegt darin, daß im Gegensatz zu Glycoltrinitrat Natriumnitroprussid und SiN-1 direkt NO freisetzen und keinen zusätzlichen, intrazellulären Konversionsschritt benötigen.

Zusammenfassungen und Schlußfolgerungen

Die wichtigsten Gesichtspunkte über die Nitroglycerinwirkung und die Interaktion mit EDRF bzw. NO zeigt die nachfolgende Übersicht. Von besonderer Bedeutung ist, daß durch Nitrovasodilatatoren möglicherweise eine therapeutische Substitution möglich ist bei einer gestörten Freisetzung von EDRF bzw. NO. Das gilt insbesondere bei Anwesenheit von geschädigten bzw. atheromatösen Endothelabschnitten. Darüber hinaus gibt es einen Synergismus zwischen EDRF bzw. NO und anderen endothelialen Lokalhormonen wie z. B. dem Prostaglandin I_2 (PGI_2), so daß es zu einer Potenzierung der Thrombozytenaggregationshemmung durch die gleichzeitige Freisetzung dieser beiden endothelabhängigen Autakoide kommen kann.

Nitroglycerinwirkungen und Interaktionen mit EDRF (NO)

Günstige therapeutische Wirkungen wahrscheinlich durch:

1) Unterschiedlich starke Wirkungen auf spezifische Gefäßabschnitte (Venen, Koronarien werden stärker dilatiert als Widerstandsgefäße). Resultat: Senkung der Vorlast, verbesserte O_2-Zufuhr
2) Ähnlichkeit mit endogenem Endothel-abhängigem Dilatator NO, mögliche Wirkung durch therapeutische Substitution
3) Synergismus mit anderen endothelialen Lokalhormonen (z. B. PGI_2): Potenzierung von antiaggregatorischen Wirkungen
4) verstärkte Wirkung von Nitrovasodilatatoren an Endothel-geschädigten (atheromatösen) und denudierten (aber noch erweiterungsfähigen) Abschnitten

Unerwünschte Nebenwirkungen: Desensibilisierung (Toleranz).

Literatur

Bassenge E, Busse R (1988) Endothelial modulation of coronary tone. Prog Cardiovasc Dis 30:349–380

Bassenge E, Heusch G (1990) Regulation of Coronary Blood Flow: Nervous, hormonal and endothelial factors. Rev Physiol Biochem Pharmacol (in press)

Bassenge E, Stewart DJ (1988) Interdependence of pharmacologically-induced and endothelium-mediated coronary vasodilatation in antianginal therapy. Cardiovasc Drugs 2:27–34

Bassenge E, Busse R, Pohl U (1989) Hemmung der Thrombozytenaggregation und -adhäsion durch EDRF und deren pathophysiologische Bedeutung (Inhibition of platelet-aggregation and -adhesion by EDRF: pathophysiological significance). Z Kardiol 78 [Suppl 6]:54–58

Busse R, Lückhoff A, Bassenge E (1987) Endothelium-derived relaxant factor inhibits platelet activation. Naunyn Schmiedebergs Arch Pharmacol 336:566–571

Busse R, Pohl U, Mülsch A, Bassenge E (1989) Modulation of the vasodilator action on SIN-1 by the endothelium. J Cardiovasc Pharmacol 14 [Suppl 11]:S81–S85

Drexler H, Zeiher AM, Wollschläger H, Meinertz T, Just H, Bonzel T (1989) Flow-dependent coronary artery dilatation in humans. Circulation 80:466–474

Furchgott RF, Zawadzki JV (1980) The obligatory role of endothelial cells in the relaxation of arterial smooth muscle by acetylcholine. Nature 288:373–376

Holtz J, Giesler M, Bassenge E (1983) Two dilatory mechanisms of anti-anginal drugs on epicardial coronary arteries in vivo: indirect, flow-dependent, endothelium-mediated dilation and direct smooth muscle relaxation. Z Kardiol 72 [Suppl 3]:98–106

Mülsch A, Busse R, Bassenge E (1988) Desensitization of guanylate cyclase in nitrate tolerance does not impair endothelium-dependent responses. Eur J Pharmacol 158:191–198

Mülsch A, Busse R, Winter I, Bassenge E (1989) Endothelium- and sydnonimine-dependent responses of native and cultured aortic smooth muscle cells are not impaired by nitroglycerin tolerance. Naunyn Schmiedebergs Arch Pharmacol 339:568–574

Mülsch A, Busse R, Bassenge E (1989) Clinical tolerance to nitroglycerin is due to impaired biotransformation of nitroglycerin and biological counterregulation, not to desensitization of guanylate cyclase. Z Kardiol 78 [Suppl 2]:22–25

Pohl U, Busse R (1989) EDRF-induced increase of cGMP in platelets during passage through the coronary vascular bed. Circ Res 65:1798–1803

Pohl U, Busse R, Kuon E, Bassenge E (1986) Pulsatile perfusion stimulates the release of endothelial autacoids. J Appl Cardiol 1:215–235

Rafflenbeul W, Bassenge E, Lichtlen P (1989) Competition between endothelium-dependent and nitrogylcerin-induced coronary vasodilation (Konkurrenz zwischen endothelabhängiger und Nitroglycerin-induzierter koronarer Vasodilatation). Z Kardiol 78 [Suppl 2]:45–47

Rubanyi GM, Romero JC, Vanhoutte PM (1986) Flow-induced release of endothelium-derived relaxing factor. Am J Physiol 250:H1145–H1149

Stewart DJ, Holtz J, Bassenge E (1987) Long-term nitroglycerin treatment: effect on direct and endothelium-mediated large coronary artery dilation in conscious dogs. Circulation 75:847–856

Zeiher AM, Drexler H, Wollschläger H, Just H (1989) Preserved flow mediated vasodilation despite acetylcholine-induced vasoconstriction in atherosclerotic coronary arteries in man (abstract). J Am Coll Cardiol 13:132A

Angina pectoris – Warum wird eine stabile Form instabil?

P. R. Lichtlen

Einleitung

Die Frage nach den Ursachen des Übergangs einer stabilen Angina pectoris in die instabile Form wurde in den letzten Jahren von zahlreichen klinischen Zentren untersucht. Die *stabile Angina pectoris* ist definiert durch einen Zustand von relativ konstant auftretenden, reversiblen, ischämischen Episoden mit oder ohne typische Schmerzen, vorwiegend ausgelöst durch Belastungen bzw. Situationen erhöhten myokardialen Sauerstoffbedarfs. Anginöse Schmerzen treten jedoch nur bei ca. 20–30% der ischämischen Episoden dieser Patienten auf, ca. 70–80% der Episoden verlaufen klinisch stumm (Hausmann et al. 1987; Nikutta et al. 1989). Bei ca. 80% der Patienten treten die asymptomatischen Episoden bei erhöhtem myokardialem Sauerstoffbedarf auf (Nikutta et al. 1989), wobei diesen Episoden in 80–90% der Fälle eine Erhöhung der Herzfrequenz vorausgeht; der Herzfrequenzanstieg beträgt einige Minuten vor dem ersten Auftreten der ST-Streckensenkung 10–15 Schläge pro Minute (Hausmann et al. 1987); ca. 20% der Episoden zeigen keinen Frequenzanstieg (Nikutta et al. 1989). Die Inzidenz der ischämischen Episoden zeigt überdies einen typischen zirkadianen Verlauf mit einem Maximum in den Morgenstunden, um ca. 10–11 Uhr und den späteren Abendstunden, ca. 17–19 Uhr, während nach Mitternacht ein Minimum erreicht wird (Hausmann et al. 1987). Die Ischämieschwelle und die Anfallshäufigkeit bleiben bei stabiler Angina pectoris in der Regel über lange Zeit weitgehend konstant.

Unter einer *instabilen Angina pectoris* versteht man als erstes eine *neu auftretende Angina,* vowiegend bei Belastung oder auch in Ruhe, wobei innerhalb von Tagen bis wenigen Wochen die Attacken an Häufigkeit, Intensität und Dauer zunehmen (Lichtlen 1980, 1985; Nellessen et al. 1986, 1988). Eine zweite, häufigere Form stellt das Auftreten einer Crescendoangina bei bereits bestehender stabiler Angina pectoris dar. Auch hier kommt es zu einer Zunahme der Häufigkeit und Intensität der Attacken, die Belastungsschwelle nimmt ab, und es treten anginöse Schmerzen in Ruhe auf (Lichtlen 1985). Bei beiden Formen sind die Anfälle in der Regel begleitet von ST-Streckensenkungen als Zeichen der Innenschichtischämie; es finden sich aber auch ST-Hebungen entsprechend einer transmuralen Ischämie, was v.a. bei spastischem Gefäßverschluß und primärer Senkung des Koronarflusses beobachtet wird (Abb. 1); (Lichtlen u. Rafflenbeul 1985; Nellessen et al. 1988b), wie dies schon

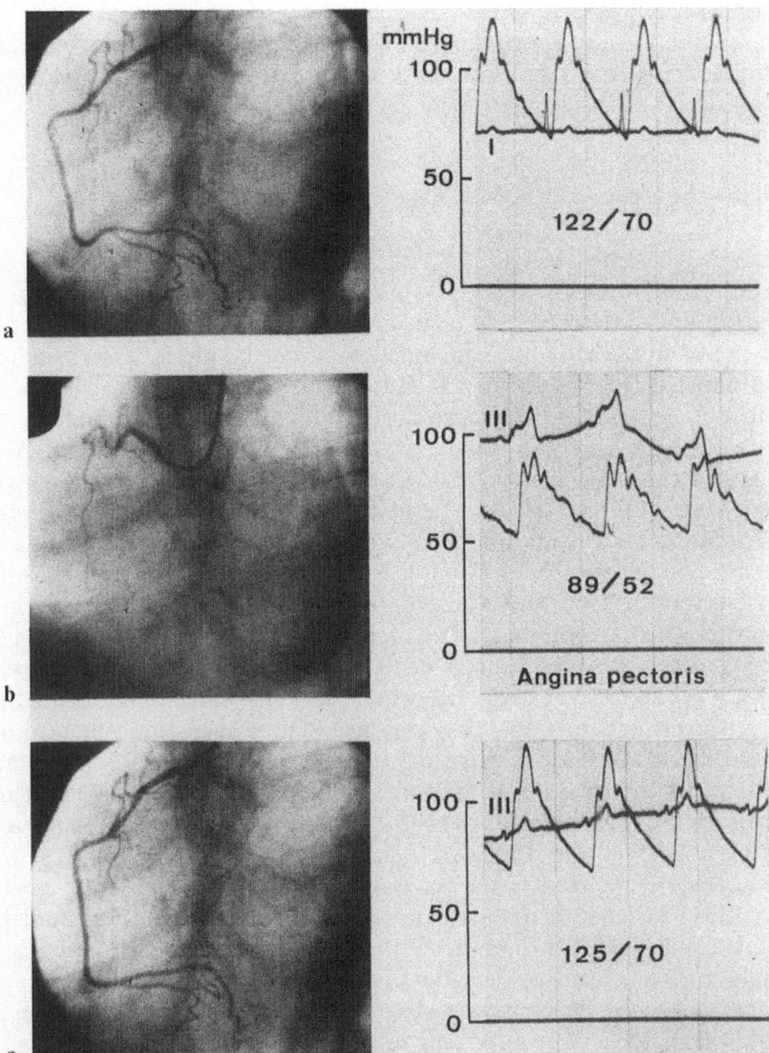

Abb. 1a–c. Typischer Spasmus im Bereich der rechten Koronararterie bei Vorliegen einer Koronarsklerose, v. a. im Bereich der linken Koronararterie (hochgradige RIVA-Stenose, Status nach Vorderwandinfarkt). Wahrscheinlich spontaner spastischer Verschluß der rechten Kranzarterie ca. 2 cm unterhalb des Ostiums mit transmuraler Ischämie (typische ST-Streckenhebung) und deutlichem Druckabfall als Ausdruck der ischämischen linksventrikulären Kontraktionseinschränkung; 5 min später spontane Erholung. **a** 1. Injektion des Kontrastmittels 8^{15} **b** 2. Injektion des Kontrastmittels 8^{20} **c** 3. Injektion des Konstrastmittels 8^{25}

von Prinzmetal et al. 1959 berichtet wurde (Prinzmetal et al. 1959; Rafflenbeul u. Lichtlen 1982). Relativ häufig finden sich im EKG spitze T-Inversionen als Ischämiefolgestadium. Ein Anstieg der kardiospezifischen Enzyme wie bei Infarkt ist definitionsgemäß nicht nachweisbar.

Ursachen der instabilen Angina pectoris

Die wesentliche Ursache des Übergangs in eine instabile Form der Angina pectoris ist in einer *raschen Zunahme des Stenosegrades eines Koronargefäßes* zu suchen, z. B. wenn eine niedriggradige, nicht ischämisch wirkende Stenose von z. B. unter 70 % bzw. mit einem minimalen Durchmesser von über 1,5 mm in eine kritische Stenose z. B. mit einem Stenosegrad von 80 % oder mehr übergeht, wobei der Gefäßdurchmesser deutlich unter 1 mm liegt. Dabei geht man heute davon aus, daß der anatomischen Komponente, z. B. der abrupten Größenzunahme der atherosklerotischen Plaque, eine wesentlich größere Bedeutung zukommt als der funktionellen Komponente, z. B. einer isolierten Erhöhung des Vasomotorentonus, einem Spasmus, z. B. aufgrund eines erhöhten α-adrenergen Tonus (Heusch et al. 1984). Der typische Verlauf stellt sich wie folgt dar: Über einen langen Zeitraum verläuft die Progression der Koronarsklerose asymptomatisch, dabei kommt es zum Auftreten von neuen, niedriggradigen Plaques (Fuster et al. 1987; Lichtlen et al. 1990). Ischämische Attacken treten erst dann auf, wenn diese sich zu einem höheren Stenosegrad entwickelt haben, wobei das weitere „Plaquewachstum“ nicht durch den eigentlichen Prozeß der Atherosklerose, sondern durch Komplikationen der Plaque initiiert wird, z. B. Blutungen in die Plaque aus Vasa vasorum, Plaquerupturen usw. (Moise et al. 1984; Ambrose et al. 1986; Fuster u. Chesebro 1986). Die schwerwiegendste Komplikation stellt die Ruptur der fibrösen Deckplatte der Plaque mit der Folge der raschen Ausbildung eines okkludierenden Plättchenthrombus dar. Dieser Mechanismus ist sicherlich einer der entscheidensten Faktoren der Entstehung der instabilen Angina pectoris. Abbildung 2 zeigt schematisch die typische Plaque mit der fibrösen Kappe, welche aus kollagenen Fasern sowie Monozyten und Makrophagen besteht (Fuster et al. 1987). Man geht aufgrund relativ großer histologischer Erfahrung (z. B. Davies u. Thomas 1984) heute davon aus, daß die Ruptur der fibrösen Kappe einer Plaque ein häufiges Ereignis darstellt, wobei dieser Vorgang nicht vom Stenosegrad abhängt bzw. auch häufig bei niedriggradigen Stenosen vorkommt und subklinisch verläuft (Ambrose et al. 1986; Nellessen et al. 1988 a, b). Die Ursachen des Einrisses der fibrösen Deckplatte sind noch unklar. Man vermutet, daß die fibröse Klappe möglicherweise durch kollagenabbauende Enzyme aufgedaut wird (Davies u. Thomas 1986); zusätzlich können Spasmen bzw. eine Steigerung des Vasomotorentonus in den noch normalen Wandanteilen dieser in der Regel exzentrischen Stenosen (Freudenberg u. Lichtlen 1981) zu einem Zug an der fibrösen Deckplatte führen und so deren Einreißen begünstigen. Eine weitere mögliche Ursache sind Blutungen innerhalb der Plaque aus neu gebildetem Vasa vasorum, was zu einer erheblichen Volumen-

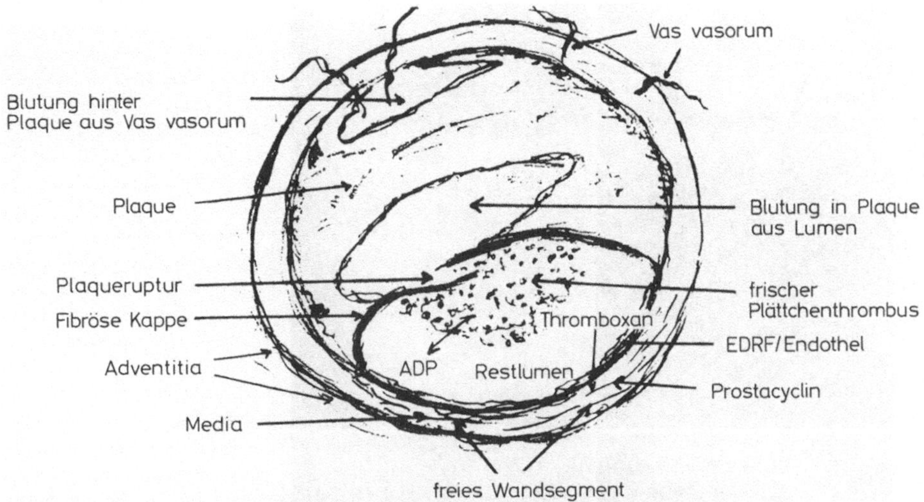

Abb. 2. Schematische Darstellung der Plaqueruptur mit ihren Folgen (s. Text)

und Druckzunahme bzw. Spannung innerhalb der Plaque und damit zu einem Einriß der Deckplatte führen kann (Fuster et al. 1988). Die klinisch entscheidende Folge eines solchen Deckplattenrisses bzw. einer Plaqueruptur ist die sofortige Anlagerung von Thrombozyten im Rupturbereich und die Ausbildung eines Plättchenthrombus, welcher bei höhergradigen Stenosen rasch zum Gefäßverschluß führt (Davies u. Thomas 1986; Fuster et al. 1988). Wahrscheinlich handelt es sich beim Einriß der Deckplatte und der Bildung eines Plättchenthrombus um ein relativ häufiges Ereignis. Bei Plaques, die das Lumen noch nicht kritisch stenosieren, treten die thrombotischen Anlagerungen klinisch nicht immer in Erscheinung (Mandelkorn et al. 1983; Fuster u. Chesebro 1986; Fuster et al. 1987). Die Thromben werden entweder spontan lysiert oder in die Plaque integriert (Fuster u. Chesebro 1986), was zu einer Zunahme des Stenosegrades bzw. einer sekundären Progression der Koronarsklerose führt. Kritisch wird die Situation, wenn bei einer schon höhergradigen Stenose (z. B. über 75 %) der an der Rupturstelle sich neu bildende Plättchenthrombus das Lumen in kurzer Zeit (Minuten) verschließt, was aufgrund der ausfallenden Sauerstoffzufuhr rasch (innerhalb von 30 min) zu einem Infarkt im abhängigen Myokardbereich führt (Schaper et al. 1985). Ferner können bei der Ablösung von Randpartien der Thromben periphere Embolisierungen entstehen, welche evtl. zu Arrhythmien bzw. zum plötzlichen Herztod führen (Davies u. Thomas 1986).

Die Ausbildung eines das Lumen nur teilweise verschließende Plättchenthrombus als Ursache der instabilen Angina ist heute gut belegt (Davies u. Thomas 1986; Davies et al. 1988; Fuster u. Chesebro 1986; Fuster et al. 1988; Nellessen et al. 1988). Als Beispiel sei ein Patient aufgeführt, der 3 Wochen nach einem frischen Herzinfarkt, kurz vor der Entlassung, eine klassische

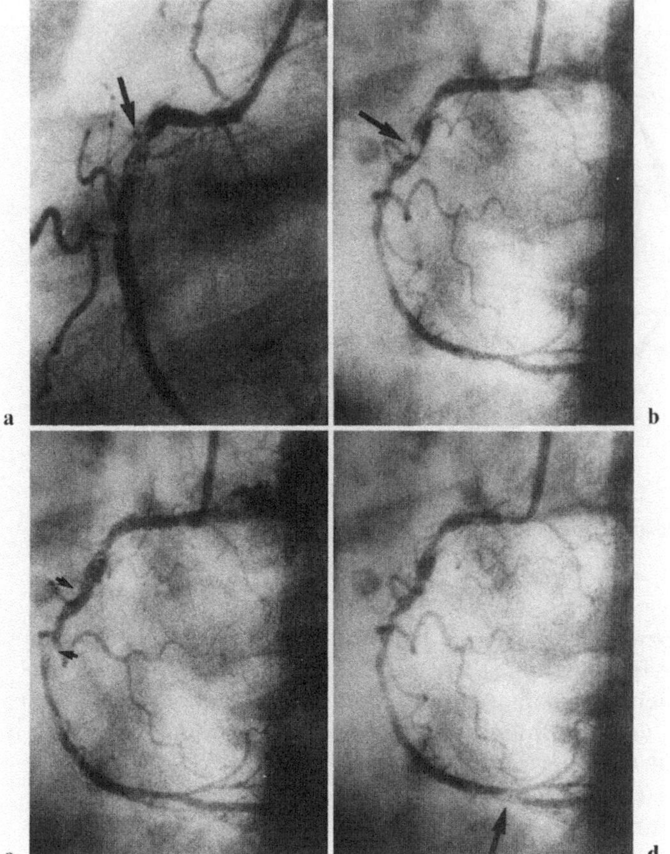

Abb. 3a–d. Ausbildung eines weitgehend okkludierenden Plättchenthrombus im Bereich einer niedriggradigen Plaque bzw. Stenose in der proximalen rechten Kranzarterie (*große Pfeile*) wahrscheinlich aufgrund einer Plaque-Ruptur; durch sofortige Thrombolyse mit Streptokinase rasche Auflösung des Thrombus mit gleichzeitiger vorübergehender peripherer Embolisierung (**d:** *Pfeil;* Einzelheiten s. Text)

instabile Angina pectoris entwickelte (Abb. 3). Bei der innerhalb einer Stunde durchgeführten Angiographie ließen sich im Bereich der rechten Kranzarterie eindeutig Thromben nachweisen, die sich an eine rupturierte Plaque angelagert hatten. Es wurde mit Erfolg eine Lysetherapie durchgeführt. Daß sich solche Formationen von Plättchenthromben bei Auftreten einer instabilen Angina nicht immer röntgenologisch beobachten lassen, ist im zeitlichen Ablauf der Ereignisse begründet; Untersuchungen, bei denen die Relation zwischen der Häufigkeit von angiographisch sichtbaren Thromben und dem Intervall zwischen Beginn der anginösen Symptome und der Angiographie verglichen wurden, haben gezeigt, daß die Wahrscheinlichkeit, einen Thrombus zu entdecken, um so größer ist, je früher nach Beginn der Symptome die Angiographie durchgeführt wird (Tabelle 1) (Nellessen et al. 1988 a). Während nach mehr als 120 Tagen – wegen der Spontanlyse – nur noch bei ca. 1 % der Patienten Thromben gefunden werden, lassen sich bei ca. 50 % der Patienten mit instabiler Angina Thromben nachweisen, wenn die Angiographie innerhalb von 2–7 Tagen durchgeführt wird. Für die Bedeutung der Plättchenthromben bei der Entstehung der instabilen Angina pectoris spricht auch die Beobachtung, daß es bei diesen Patienten zu einem relativ hohen Prozentsatz (10–15 %) innerhalb kurzer Zeit (in Tagen nach Schmerzbeginn) trotz entsprechender Therapie zu Myokardinfarkten kommt, wobei stets ein Gefäßverschluß durch einen Thrombus gefunden wird (Tabelle 2) (Lichtlen 1985 a; Davies u. Thomas 1986; Nellessen et al. 1988).

Ein weiterer, wesentlicher Faktor liegt darin, daß die Plättchenthromben z. T. in die Plaques integriert werden und damit zur Progression der Koronarsklerose beitragen (Davies u. Thomas 1986; Fuster et al. 1987). Werden Angio-

Tabelle 1. Beziehung (Δt) zwischen Zeitpunkt der Angiographie und Bestehen von Thromben bei Patienten mit instabiler Angina pectoris

Autoren	*n*	Häufigkeit		Δt (Tage)
		n	[%]	
Holmes	1202	16	(1)	<120
Mandelkorn	9	4	(44)	<6
Vetrovec	129	8	(6)	<30
Capone I	119	44	(37)	<14
Capone II	44	23	(52)	<2
Cowley	80	40	(50)	<2
Rafflenbeul	52	22	(42)	<7

Tabelle 2. Infarktinzidenz bei instabiler Angina pectoris (Literaturübersicht). (*HMS* Studie der Abteilung für Kardiologie der Medizinischen Hochschule Hannover). (Nach Nellessen et al. 1988)

Autoren	Jahr	*n*	Hospitalisierte Patienten [%]	Nach 1 Jahr[a] [%]
Krauss	1972	100	8	13
Gazes	1973	140	17	–
Duncan	1976	251	7	13 (6 Monate)
Heng	1976	158	13	–
US Coop	1978	288	12,5	11,8
Russell	1978	147	8[b]	–
Mulcahy	1981	101	9[b]	12[b]
Lewis	1983	641	–	7 (2 Monate)
HMS	1986	123	11,4	12,2

[a] Außer Todesfälle im Krankenhaus
[b] Nur nichttödlich verlaufene Infarkte

gramme in Intervallen durchgeführt, so läßt sich eine Progression wesentlich häufiger bei Patienten mit instabiler als mit stabiler Angina pectoris beobachten. Zwei unabhängige Arbeitsgruppen (Moise et al. 1984; Ambrose et al. 1985) konnten nachweisen, daß bei Patienten mit instabiler Angina pectoris, welche 35 bzw. 27 Monate nach einer Erstangiographie auftrat, in 76% bzw. 59% der Fälle neue Stenosen und in 50% und 45% sogar neue Verschlüsse aufgetreten waren. Bei Patienten mit stabiler Angina pectoris war hingegen die Inzidenz neuer Stenosen mit 31% und 18% wesentlich geringer; dies entspricht auch der Erfahrung aus retrospektiven Intervallstudien, wo eine jährliche Progression bei ca. 20% der Patienten mit stabiler Angina pectoris gefunden wird (Nellessen et al. 1988a).

Ein weiterer wichtiger Gesichtspunkt betrifft die *Morphologie der Stenosen*. Ausgedehnte angiographische Studien haben ergeben, daß gewisse Stenosen z.T. sehr unregelmäßige Begrenzungen, Rezessi und Nischen aufweisen, in denen sich thrombotisches Material anlagert. Diese sog. „komplizierten“ Stenosen (Abb. 4) mit rauher Oberfläche und kantigen Konturen findet man relativ häufig bei instabiler Angina pectoris oder Myokardinfarkt (Ambrose et al. 1986), häufiger als sog. unkomplizierte Stenosen, die glatt begrenzt sind bzw. eine glatte Oberfläche aufweisen können (Abb. 5). Unkomplizierte Stenosen lassen sich dementsprechend auch über Jahre als unverändert nachweisen (Lichtlen 1985a).

Als letzter der eine Entstehung der instabilen Angina pectoris begünstigenden Faktoren ist die *funktionelle Komponente* zu erwähnen. Postmortale, histologische Untersuchungen (Freudenberg u. Lichtlen 1981) haben gezeigt, daß ein Großteil der Koronarstenosen in bezug auf die Gefäßzirkumferenz exzentrisch lokalisiert ist. 75% der atherosklerotischen Plaques nehmen nur einen Teil der Gefäßwand ein bzw. sind exzentrischer Natur, während der übrige Wandanteil noch ein histologisch weitgehend normales Gewebe aufweist. Die glatte Gefäßmuskulatur dieses Wandsegments ist noch in der Lage, sich zu kontrahieren und zu dilatieren und damit den Stenosegrad rasch zu verändern, im Sinn einer weiteren Zu- oder Abnahme, was zum Begriff der sog. „dynamischen“ Stenose geführt hat (Rafflenbeul u. Lichtlen 1978). Exzentrische und konzentrische Stenosen (die letzteren nehmen die gesamte Gefäßzirkumferenz ein) lassen sich angiographisch leicht unterscheiden, da bei exzentrischen Stenosen der minimale Durchmesser z. B. unter Nitraten in verschiedenen Projektionen erheblich variiert, während er bei konzentrischen weitgehend unverändert bleibt (Lichtlen u. Rafflenbeul 1985). Die Unterscheidung zwischen exzentrischer und konzentrischer Stenose hat für die Therapie erhebliche Bedeutung (Lichtlen 1985b). Bei einigen Patienten mit schwerer instabiler Angina pectoris ist – wegen der spastischen Einengung – eine angiographische Darstellung der betroffenen Koronargefäße nur nach Verabreichung von massiven Dosen von Nitraten bzw. Kalziumantagonisten, also die glatte Gefäßmuskulatur relaxierenden Substanzen, möglich. Hier liegen in der Regel exzentrische Stenosen vor, die unter vasorelaxierenden Substanzen (Nitrate und/oder Kalziumantagonisten) häufig noch deutlich, sogar bis in den subklinischen Bereich, erweitert werden können (Abb. 5) (Rafflenbeul 1980; Nelles-

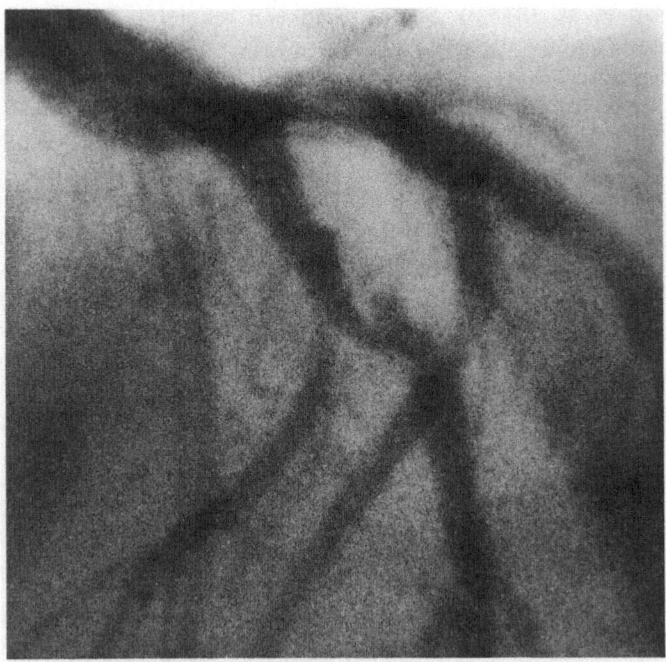

Abb. 4. Typische komplizierte Stenose im Bereich des proximalen RIVA; die Plaque ragt distal dachziegelartig in das Gefäß hinein, wodurch eine typische Nische entsteht (solche Formen finden sich v.a. nach Plaqueruptur)

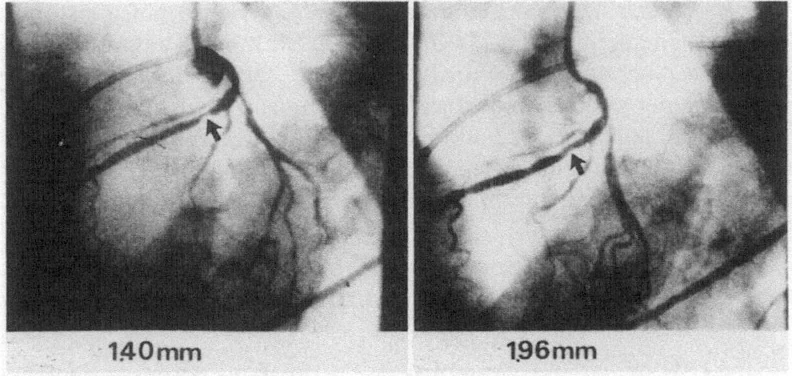

Abb. 5. Typisches Beispiel einer unkomplizierten, exzentrischen, dynamischen Stenose im proximalen Abschnitt des RIVA (*Pfeil*). Unter Verabreichung von 0,8 mg Nitroglycerin sublingual und 20 mg Nifedipin oral deutliche Erweiterung des minimalen Stenosedurchmessers von im Mittel 1,40 auf 1,96 mm (um ca. 40 %) aufgrund der Relaxation des restlichen noch normalen Gefäßwandabschnitts im Stenosebereich

sen et al. 1988 b). Solche Stenosen reagieren umgekehrt aber auch relativ häufig auf Einflüsse, welche den lokalen Vasomotorentonus erhöhen bzw. zu Spasmen führen. Dabei kann es sich um ein Überwiegen der α-adrenergen Innervation handeln (Heusch et al. 1984), um eine Schädigung der Endothelfunktion mit der Folge einer verminderten oder fehlenden Freisetzung von EDRF, dem endothelial relaxierenden Faktor (Busse u. Bassenge 1985; Bossaller et al. 1987; Mügge et al. 1989) oder um vasoaktive Substanzen, die bei der Bildung eines Plättchenthrombus freigesetzt weren (Thrombin, ADP) (Mügge et al. 1989). Das Fehlen von EDRF aufgrund eines Endothelschadens (Endothel gänzlich fehlend oder neu gebildet und funktionslos) im Bereich atherosklerotischer Plaques (Davies et al. 1988) spielt möglicherweise bei der instabilen Angina pectoris eine wesentliche Rolle; die Bedeutung dieses Faktors wird zur Zeit weiter geklärt. Alle diese verschiedenen pathogenetischen Mechanismen können in dynamischen Stenosen Koronarspasmen hervorrufen und eine instabile Angina pectoris auslösen.

Zusammenfassung und Schlußfolgerungen

Der Übergang einer stabilen Form der Angina pectoris in eine instabile beruht auf dem Zusammenspiel mehrerer Faktoren. Im Vordergrund stehen anatomische Veränderungen, ein rasches „Wachstum" der atherosklerotischen Plaque – als Folge einer Ruptur der Deckplatte – mit Anlagerung eines evtl. okkludierenden Plättchenthrombus. Als zusätzlicher wichtiger Faktor ist die funktionelle Komponente bei exzentrischen Stenosen anzusehen, wo noch ein normales Wandsegment vorliegt, das sich sowohl zu kontrahieren wie zu relaxieren vermag (dynamische Stenose). Schließlich dürfte das im Bereich der atherosklerotischen Plaque häufig fehlende oder neu gebildete funktionslose Endothel durch Mangel an EDRF zur Steigerung des Vasomotorentonus im Stenosebereich beitragen und damit dem Entstehen einer instabilen Angina pectoris Vorschub leisten. Insgesamt gesehen führen daher überwiegend anatomische, aber auch funktionelle Momente, evtl. beide gemeinsam, zum Übergang von der stabilen in die instabile Form der Angina pectoris, welche wegen der zugrundeliegenden anatomischen Änderungen ein prognostisch ungünstiges Stadium der koronaren Herzkrankheit darstellt – mit relativ häufigem Übergang in einen Herzinfarkt oder plötzlichen Herztod.

Literatur

Ambrose JA, Winters SL, Arora RR, Eng A, Riccio A, Fuster V (1985) Evaluation of coronary morphology in patients progressing to unstable angina (abstract). J Am Coll Cardiol, p 519

Ambrose JA, Winters SL, Rohit RA, Eng A, Riccio A, Gorlin R, Fuster V (1986) Angiographic evaluation of coronary artery morphology in unstable angina. J Am Coll Cardiol 7/3:472–478

Bossaller C, Yamamoto H, Henry PD, Lichtlen PR (1987) Regulation des Tonus menschlicher Koronargefäße über einen endothelialen relaxierenden Faktor. Z Herz Thorax Gefäßchir 1:62–66
Busse R, Bassenge E (1985) Regulation des Gefäßtonus über das Endothel. Z Kardiol [Suppl 7] 74:99–106
Davies MJ, Thomas A (1984) Thrombosis and acute coronary artery lesion in sudden ischemic death. N Engl J Med 310:1137–1140
Davies MJ, Thomas AC (1986) Plaque fissuring: the course of acute myocardial infarction, sudden ischemic death and crescendo angina. Br Heart J 53:363–393
Davies MJ, Woolf N, Rowles PM, Pepper J (1988) Morphology of the endothelium over atherosclerotic plaques in human coronary arteries. Br Heart J 60:459–464
Freudenberg H, Lichtlen PR (1981) Das normale Wandsegment bei Koronarstenose – eine postmortale Studie. Z Kardiol 70:863–869
Fuster V, Chesebro JH (1986) Mechanisms of unstable angina. N Engl J Med 315:1023–1025
Fuster V, Adams PhC, Badimon JJ, Chesebro JH (1987) Platelet-inhibitor drug's role in coronary artery disease. Prog Cardiovas Dis 29:325–346
Fuster V, Badimon L, Cohn M, Ambrose JA, Badimon JJ, Chesebro J (1988) Insights into the pathogenesis of acute ischemic syndromes. Circulation 77:1213–1220
Hausmann D, Nikutta P, Hartwig CA, Daniel WG, Lichtlen PR (1987) ST-segment analysis in the 24-h Holter ECG in patients with stable angina pectoris and proven coronary artery disease. Z Kardiol 76:554–562
Heusch G, Deussen A, Schipke Z, Thämer V (1984) Alpha-1- and alpha-2-adrenoceptor-mediated vasoconstriction of large and small canine coronary arteries in vivo. J Cardiovasc Pharmacol 6:961–968
Lichtlen PR (1980) Klinik, Diagnostik und Therapie der unstabilen Angina pectoris. Internist (Berlin) 21:636–645
Lichtlen PR (1985a) Unstabile angina pectoris – Status 1985. Hämostasiologie 6:102–122
Lichtlen PR (1985b) Pathophysiology of coronary and myocardial function in angina pectoris: important aspects for drug treatment. Eur Heart J [Suppl F] 6:11–25
Lichtlen PR, Rafflenbeul W (1985) Effects of calcium antagonists on fixed and dynamic obstructions in patients with severe coronary artery disease. In: Fleckenstein A, Breemen C van, Gross R, Hoffmeister F (eds) Cardiovascular effects of dihydropyridine-type calcium antagonists and agonists. Springer, Berlin Heidelberg New York Tokyo, pp 381–407
Lichtlen PR, Hugenholtz PG, Rafflenbeul W, Hecker H, Jost S, Nikutta P, Decker JW and the INTACT group (in preparation) Retardation of coronary artery disease in man by the calcium channel blocker nifedipine; results of INTACT. Cardiovasc Drugs Ther
Mandelkorn JB, Wolf NM, Singh S (1983) Intracoronary thrombus in nontransmural myocardial infarctions and in unstable angina pectoris. Am J Cardiol 52:1–6
Moise A, Théroux P, Taeymans J, Waters DD, Lesperance J, Fines P, Descoings B, Robert P (1984) Clinical and angiographic factors associated with progression of coronary artery disease. J Am Coll Cardiol 3:659–667
Mügge A, Bode S, Wahlers T, Haverich A, Lichtlen PR, Frölich JC, Förstermann U (1989a) Aggregierende menschliche Thrombozyten stimulieren die Produktion des endothelialen, relaxierenden Faktors (EDRF) in menschlichen Koronararterien. Cor Vasa 1/2:7–11
Mügge A, Förstermann U, Lichtlen PR (1989b) Endotheliale Funktionen bei kardiovaskulären Erkrankungen. Z Kardiol 78:147–160
Nellessen U, Hecker H, Danciu V, Specht S, Lichtlen PR, Borst HG (1986) Instabile Angina pectoris: Krankheitsbild und Verlauf neu überprüft. Z Kardiol 75:707–718
Nellessen U, Hecker H, Lichtlen P (1988a) Instabile Angina pectoris. Klinik, Therapie und Prognose. Thieme, Stuttgart New York
Nellessen U, Rafflenbeul W, Daniel WG, Jost S, Hecker H, Lichtlen PR (1988b) Effects of nifedipine and nitrates on coronary vasomotion. Eur Heart J 9:83–88
Nikutta P, Hausmann D, Daniel WG, Wenzlaff P, Lichtlen PR (1989) Different circadian rhythm in heart rate dependent and independent ischemic episodes. Eur Heart J [Suppl] 10:111

Prinzmetal M, Kennamer R, Merliss R, Wada T, Bor N (1959) Angina pectoris. I. A variant form of angina pectoris. Preliminary report. Am J Med 27:375–388

Rafflenbeul W, Lichtlen PR (1982) Zum Konzept der „dynamischen" Koronarstenose. Z Kardiol 71:439–444

Rafflenbeul W, Urthaler F, Russell R, Lichtlen P, James TN (1980) Dilatation of coronary artery stenoses after isosorbide dinitrate in man. Br Heart J 43:546–549

Rafflenbeul W, Berger C, Jost S, Lichtlen PR (1987) Constriction of coronary arteries and stenoses with propranolol. Circulation [Suppl 4] 76:276

Schaper J, Alpers P, Gottwick M, Schaper W (1985) Ultrastructural characteristics of regional ischemia and infarction in the canine heart. Eur Heart J 6:21–31

New Results in Coronary Angioscopy

W. S. Grundfest, C. Beeder, J. Segalowitz, and F. Litvack

Introduction

With angioscopy, direct visualization of the interior surface of blood vessels is possible. This article will describe the principles of angioscopy and discuss the insights into pathophysiology and pre- and postinterventional states provided by this method.

Principles of Angioscopy

All angioscopes in common use today consist of a long shaft, the distal end of which has an outer plastic jacket surrounded by a series of illumination bundles. A lens at the distal end of the angioscope focuses light on a series of coherent bundles that transfer information back to the viewer.

Construction of angioscopic devices is a sophisticated and rapidly progressing field. As little as two decades ago, a modified 0.8 cm colonoscope represented the state-of-the-art in vascular endoscopy (Greenstone et al. 1966). However, the large endoscope size and its inability to keep the imaging field clear of blood limited its practical application. Over the past 6 years, we have investigated more than 50 different endoscopes (Grundfest et al. 1985; Litvack et al. 1984). A number of 0.7 to 0.5 mm devices are now available from a variety of companies, including Olympus (Rye, NY), VascuCare (Orangeburg, NY), Baxter-Edwards (Irvine, CA) and Intramed (San Diego, CA).

Improvements in fiber optics have been largely responsible for ever smaller diameter flexible endoscopes. While each individual fiber used to be as much as 12 μm in diameter, current fibers can be as small as 1.6 μm in diameter. Still, since each fiber represents one pixel in the subsequent image, the fibers need to be put together as a bundle with a lens at each end. Although the resultant fiber bundle is flexible, individual fibers can break fairly easily. As the number of broken fibers increases, image quality deteriorates, both in terms of resolution and contrast.

Angioscopic Equipment

We have developed a 5 French angioscope that fits inside an 8 French guiding catheter and over a .014 inch guide wire. The system is flushed using an unoxygenated Ringer's lactate solution at 37 °C. We initially used a cold saline solution, but switched after the saline solution unfortunately led to an episode of ventricular tachycardia in one of our patients.

The angioscopy system comprises a subselective guiding catheter, a multiple adapter and a very thin angioscope. The angioscope is 0.5 mm in diameter with a radiopaque gold band at the end. Although device delivery is easy, inability to steer the endoscope off the wall often prevents visualization of the entire lumen.

Angioscopy in Patients with Coronary Heart Disease

Over the past 4 years we have performed a number of angioscopic examinations in an attempt to characterize the angioscopic morphology of acute coronary syndrome. We looked at a variety of patients: 18 with stable angina, 12 with progressive angina, 17 with rest angina, 15 postmyocardial infarction and two postthrombolysis patients in the intraoperative setting.

A system was developed to characterize our observations. A vessel was referred to as stable if the surface was smooth or yellow-white. An angioscopically irregular, fissured or hemorrhagic blood-stained surface represented ulcerated plaque. A red coalescence of blood that did not disappear, despite flushing irrigating fluid over it, was a thrombus. This classification system was tested against in vitro studies of human coronary arteries and grafts, as well as animal arteries, and appears to be valid.

A typical stable atheroma is shown in Figure 1 A. It is smooth and white, and there is little or no irregularity in the surface. When you look at this surface with scanning electron microscopy you see that it is relatively well organized, with all of the nuclei aligned in the same direction, an intact intimal surface and smooth intimal cells. However, in an ulcerated surface (Fig. 1 B), one sees a small hemorrhage and an irregular, eccentric lumen with a rough, irregular surface. This ulcer was induced in an interesting subject: a previously non-smoking rat who was induced to try its first (and last) cigarette. After the cigarette, the rat's coronary artery showed a number of these ulcers, with platelets, red blood cells and fibrin debris attached very firmly to an ulcerated surface.

The exact same pattern is seen in patients with an ulcerated coronary artery. After the ulceration the vessel wall becomes a nidus for thrombus formation on the surface (Fig. 1 C). These red masses that cannot be "washed" away have a typical appearance, can be partially obstructing and eventually become circumferential. If the patient goes on to total occlusion and infarction, one sees a lumen totally occupied by thrombus.

In Figure 2 we see a reoccluded artery in a patient 24 hours post thrombolysis with tissue plasminogen activator. After significant irrigation for approximately 4 minutes, the reason why the artery could not be kept open became apparent (Fig. 2B). There is an ulcerated plaque, a series of laminated or colored parts of the wall with the incorporation of thrombus. Post irrigation we see the ulcerated plaque and layers of thrombus and a very narrowed lumen (Fig. 2C).

Figure 3 shows a typical angioscopic view of an artery in which acute percutaneous transluminal coronary angioplasty (PTCA) has failed. What is probably white thrombus or debris is seen filling the artery. This was an acute case and with no time for trombus formation. Histologically, a ruptured arterial surface is seen with suffusion of blood into the underlying layers in between the intima and media, along with disruption of the intimal surface.

Insights in Pathophysiology

We have learned from our studies that a coronary thrombosis is dynamic. It evolves by layering, lysis or incorporation: it is not a fixed entity. In fact, the key to understanding coronary artery disease is realizing how rapidly it can change.

We believe that rapid progression of isolated coronary stenosis is probably caused by organization and incorporation of thrombi. We have seen that a balloon angioplasty can be performed without intimal disruption. But when balloon angioplasty fails, the failure is almost always caused by dissection and associated thrombus, which are not always angiographically detected. In our series of 20 patients who underwent angioscopy in failed PTCA, 15 had lesions that were not angiographically detected. The lesions escaped detection either at the time PTCA was attempted or acutely, when the patient returned for a repeat angiogram or underwent surgery.

Table 1 shows a summary of acute and chronic syndromes and a distribution of endothelial abnormalities. Almost all of the stable patients have had what we previously defined as stable atherosclerosis. Interestingly, one patient

Table 1. Acute and chronic coronary syndromes: Distribution of endothelial abnormalities

Anginal syndrome	Stable atherosclerosis (*n*)	Ulcerated plaque (*n*)	Thrombus (*n*)
Stable	17/18	1/18	0/18
Progressive	2/12	10/12	0/12
Rest	2/17	3/17	12/17
Post MI	2/15	10/15	3/15
Thrombolysis	0/2	0/2	2/2

MI, myocardial infarction

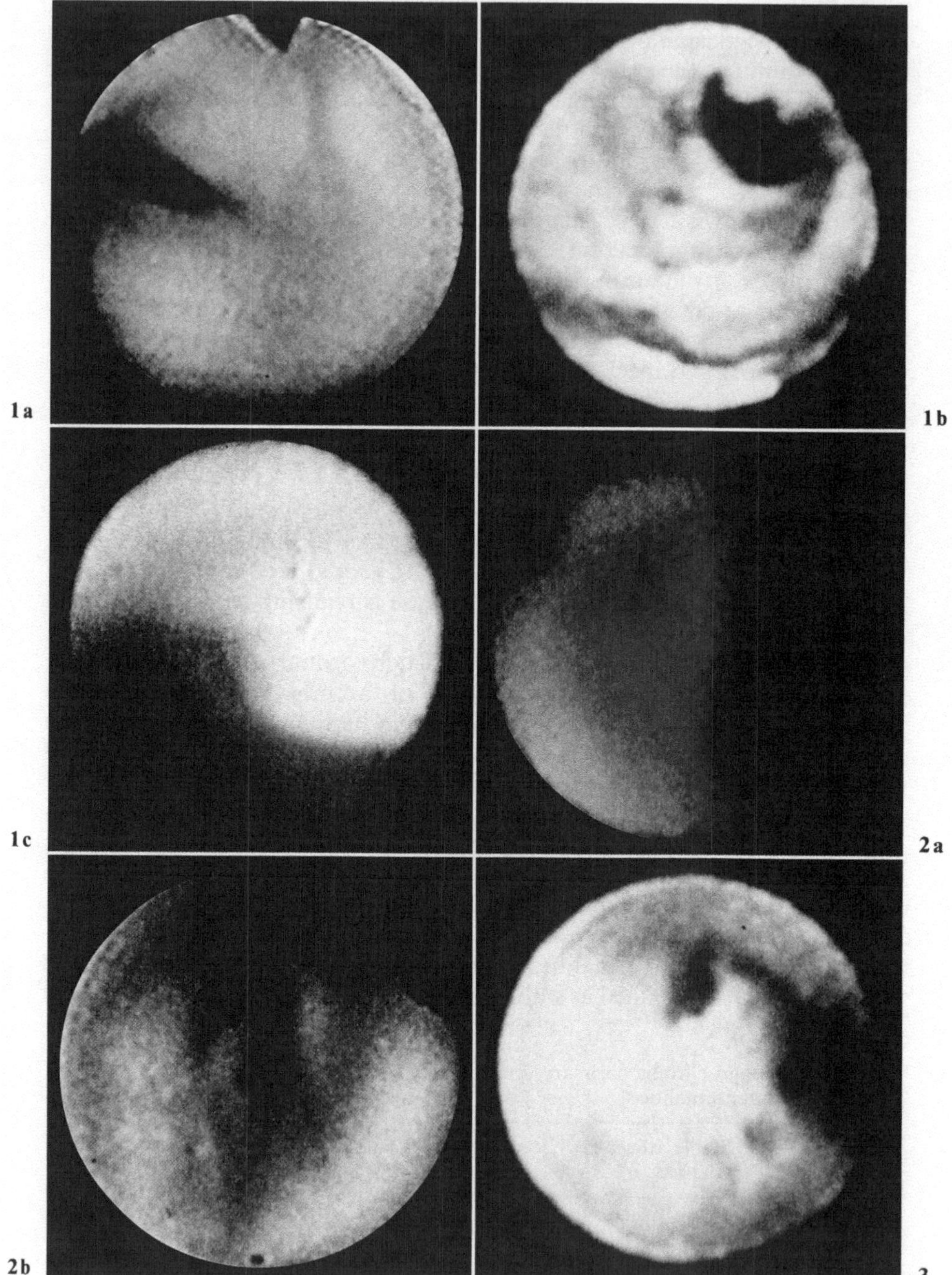
1a
1b
1c
2a
2b
3

who subsequently underwent bypass surgery and has a prior history of very stable angina induced at a certain level by exercise (approximately one flight of stairs) did have an ulcerated plaque in his LAD. However, we never saw a thrombus.

Most of the patients with progressive or accelerated angina have ulcerated plaques. Even though the most critical lesion was never identified, a thrombus was never seen in patients with accelerated angina. One might think that the distinction between accelerated and rest angina is not all that great; surprisingly, the patients with rest angina almost always had a thrombus, and the distinction between the two groups was statistically significant.

The majority of patients with rest angina had a thrombus. A few patients presented only with ulcerated plaques: presumably the thrombus had lysed. In a few patients no lesion was found. In nearly all of our post myocardial infarction patients we found an ulcerated plaque with thrombus. To date, there have only been two patients in whom we did not see the lesion. The reason for this may be that it is not always possible, especially intraoperatively, to place the angioscope optimally. These data reinforce the concept of a dynamic process. Lysis, vessel wall repair, thrombus incorporation and ulceration are all part of the cycle of atherosclerotic cardiovascular disease.

Fig. 1a–c. Endoscopic picture obtained from **a** the lumen of a patient with chronic stable angina who underwent bypass surgery for high-risk multiple vessel disease. The image shows a typical atherosclerotic plaque: smooth, yellow-white, and without ulceration or thrombus formation. **b** Left anterior descending coronary artery in a patient with typical accelerated angina with 2 weeks of increasing chest pain. The angioscopic image of the arteries shows a dramatic semicircumferential ulcer with suffused hemorrhage between the intima and media, and associated platelet debris. This image is typical for patients with accelerated angina. **c** LAD of a patient with 4½ days of unstable rest angina, unresponsive to nitrates and beta blockers, who came to bypass surgery after medical therapy failed. We see a bright red crescent-shaped thrombus in the artery. Vigorous irrigation failed to remove the thrombus, which proved to be a gelatinous mass adherent to the vessel wall. Note that the rest of the arterial surface is smooth and white

Fig. 2a–c. A dramatic example of an artery's appearance in a failed case of thrombolysis. The patient previously received TPA for recanalization of the artery after an acute MI was confirmed by EKG and enzyme evidence. Twenty-four hours after the patient's initial TPA dose the artery spontaneously reoccluded while he was in the intensive care unit, with a marked drop in blood pressure requiring urgent intervention. Angioscopy revealed an artery totally occluded by thrombus. **b** The same artery after vigorous irrigation, in which most of the thrombus has been removed from the arterial wall. **c** Underneath the thrombus we see an extremely small lumen at the 6 o'clock position, and ulcerated plaque and areas of layered thrombus. We know this is layered thrombus because of the color changes, which tend to correspond to the age of the thrombus

Fig. 3. A typical angioscopy view after a failed PTCA. Initially, in the catheterization laboratory, this appeared on angiography to be successful. However, 24 h post intervention the patient's symptoms returned, requiring urgent intervention. Angioscopy showed a large intimal flap-associated hemorrhage

From these data we can advance a concept of an ulceration-thrombosis-cycle in coronary artery disease. One assumes that there is a long-standing, progressive, slowly developing stable atheroma. For some as yet unknown reason, the stable atheroma ulcerates. The ulceration could be due to smoking, high blood pressure, or any of a number of postulated causes, including viruses or high blood levels of LDL.

This ulceration leads to several processes. Accelerated angina may develop with the accumulation of platelet debris and arterial spasm induced by the compounds released when platelets aggregate. The platelet aggregation can be dramatic, resulting in sudden death, or it can be slow and insidious, resulting in microemboli and ischemic cardiomyopathy. If the ulcer heals, we see a stenosis and progression of the stenosis with a fibrotic artery and a healed ulcer. Alternatively, this ulcer can lead not only to platelet aggregation but also to activation of the clotting cascade, in which we see partial thrombosis leading to unstable angina. Should the thrombus become occlusive and there are no collaterals, one develops a myocardial infarction or sudden death if the occlusion is big enough. On the other hand, if the occlusion is only partial, one can see progression of stenosis and thrombus incorporation or recanalization leading again to stable atheroma.

We believe that coronary angioscopy has led to the following understanding: each acute coronary syndrome is closely associated with a specific morphology (Forrester et al. 1988). Acute coronary insufficiency is always associated with ulceration and thrombus. As such, acute changes in coronary symptomatology must be dealt with aggressively, and therapy must include both antiplatelet agents and thrombolytic agents. We also believe that the choice of appropriate intervention in an acute event may be guided by a knowledge of morphology. In the future, intimal flaps and thrombi not detected by angiography may be seen by angioscopy and dealt with appropriately.

References

Forrester JS, Litvack F, Grundfest W, Hickey A (1988) Definition, pathogenesis and therapy of acute coronary syndromes in man. Prog Crit Care Med 3:5–15

Greenstone SM, Shore JM, Heringman EC et al. (1966) Arterial endoscopy (arterioscopy). Arch Surg 93:81–83

Grundfest WS, Litvack F, Sherman T et al. (1985) Angioscopy for intraoperative delineation of peripheral and coronary vascular anatomy. (American Surgical Association Abstracts, March 28)

Litvack F, Grundfest WS, Forrester J (1984) New therapy update, angioscope. Cardiovasc Rev Rep 5:790–795

Silent Myocardial Ischemia

S. P. Glasser and D. K. Arnett

Patients with coronary heart disease and angina pectoris exhibit frequent episodes of asymptomatic myocardial ischemia manifested by ST segment depression during ambulatory monitoring (Cohn 1985; Kawanishi and Rahimtoola 1987; Maseri 1987; Rozanski and Berman 1987). Despite the presumed significance of this problem, little is known about the prognostic importance of asymptomatic ischemia, risk factors associated with it, noninvasive parameters that might be predictive of its occurrence, or treatment effects, not only as it relates to a reduction in the frequency and/or extent of ischemia, but most importantly as to whether it alters the natural history.

The popularization of silent (asymptomatic) myocardial ischemia (SMI) began in the 1970s, predominantly the result of three studies. Stern and Tzivoni (1974) performed ambulatory electrocardiographic monitoring (AECG) in a group of normal persons and in a group of patients with known coronary artery disease (CAD) and found that 54% of the patients had episodes of asymptomatic ST segment depression compared to 0% in the normals (Table 1). Stern and Tzivoni (1974) also assessed the importance of these asymptomatic episodes of ST segment depression and found that the cardiac morbid event rate at 1 year was higher in patients with coronary disease who had those changes compared to those who did not. Schang and Pepine (1977) performed extensive ambulatory monitoring in a group of 20 patients with documented coronary disease and observed that 75% of all episodes of ST segment depression were unassociated with symptoms and that most of those episodes occurred at relatively low activity levels. Further, in a subgroup of these patients, ambulatory monitoring was performed during hourly sublingual nitroglycerin versus placebo. They found sixfold fewer episodes of ST segment depression in the active treatment group. Since then, SMI has been shown to be common in asymptomatic patients with CAD (Table 2; Erikssen et al. 1976; Froelicher

Table 1. Holter monitor for silent myocardial ischemia (from Stern and Tzivoni 1974)

	n	HM ST (%)
Normals	42	0
CAD	104	76 (54.3%)

CAD, coronary artery disease

Table 2. Totally asymptomatic persons with positive exercise tests

Study	*n*	*n* with CAD	F/U	Cardiac events	Mortality
Erikssen et al. (1976)	105	69	8 years	18	3
Froelicher et al. (1974)	111	34	4 years	20	2
Langou et al. (1980)*	16	12	3 years	4	0
Borer et al. (1975)	30	11	–	–	–

CAD, coronary artery disease; F/U, follow up
* Asymptomatic type II hyperlipoproteinemic subjects

et al. 1974; Langou et al. 1980; Borer et al. 1975), patients with chronic stable angina pectoris (Koehn et al., 1989; Rocco et al. 1988), unstable angina pectoris (Gottlieb et al. 1986), and patients after acute myocardial infarction (Theroux et al. 1979). It should be emphasized that silent ischemia is defined by a number of different techniques, including resting ECG, exercise ECG, AECG, nuclear studies, positron emission tomography (PET), etc. It may be that similarities exist between the techniques in terms of treatment and prognosis, but it may also be that differences exist. This must be kept in mind as the field evolves. However, long-term (24 h) ambulatory monitoring has been used extensively to quantitate ischemia by measuring ST segment depression. Ambulatory monitoring has been validated, both by continuous frequency and amplitude modulated electrocardiographic recordings in concurrence with hemodynamic observations, as an accurate method of measuring transient ST segment depression characteristic of myocardial ischemia. However, asymptomatic ischemia diagnosed by PET has demonstrated that ischemic disturbances of regional perfusion last three to five times longer than ST segment depression (Imperi and Pepine 1986; Nabel et al. 1987). Nonetheless, the benefits of using ambulatory monitoring are many. The monitoring is done in the patient's usual daily environment. The patient can therefore be monitored for 24 h or more and both symptomatic and asymptomatic episodes of ischemia can be detected and quantified according to frequency, duration, and magnitude (Sokolow and McIlroy 1986).

Asymptomatic myocardial ischemia is defined as the objective evidence of ischemia in the absence of angina or equivalent symptoms, occurring in persons with documented CAD (Pepine 1986). In the stable, symptomatic patient, the percentage of ischemic episodes which are asymptomatic, occurring during long-term ambulatory monitoring, has been reported to be 50%–80% (Maseri 1987). In a study by Imperi and Pepine (1986) of 27 patients with effort angina, 21 had episodes of asymptomatic ischemia, with 412 episodes of ischemia in a total of 2736 h of monitoring, and 50% of the episodes were asymptomatic. However, it is also known that a great deal of variability exists in the number of daily asymptomatic ischemic episodes in individual patients (Deanfield et al. 1983; Nabel 1989). In addition, subjects with ST segment depression during exercise testing have been shown to have a range (1–25 episodes) of asymptomatic ischemia per day, while asymptomatic ischemia is rare in subjects with a negative ST segment response. The duration of asymptomatic ischemia is variable, but has been found not infrequently to be prolonged and associated with marked ST segment depression. There also appears to be a strong circadian rhythm or perhaps more appropriately "arousal phenomenon" in asymptomatic ischemia which follows the heart rate, blood pressure, and catecholamine circadian pattern, with most episodes occurring between 6 a.m. and 12 noon. This increased density of ischemic events in the morning does not seem to be related to the heart rate threshold or activity level, but may be related to arousal (Rosso et al. 1987; Nabel et al. 1987).

Mechanism of SMI

The task of determining what places the angina patient at risk for developing asymptomatic ischemia is difficult because of the multifarious pathophysiologic mechanisms of ischemia. One pathophysiologic mechanism of asymptomatic ischemia appears to be an increase in coronary vasomotor tone resulting in alterations in coronary blood flow. This is somewhat different than the predominant pathophysiologic trigger of symptomatic ischemia, namely increased myocardial oxygen demand (Glasser 1987; Maseri 1987). Because of this difference in mechanisms, heart rates at which asymptomatic ischemia usually occurs are significantly lower than exercise heart rates. Additionally, episodes of asymptomatic ischemia have often been found to occur during periods of emotional or mental stress. In a recent study by Deanfield et al. (1984), it was demonstrated that simple mental stress in patients with chronic symptomatic CAD could produce ischemia not associated with anginal symptoms. Other investigators have supported this finding (Specchia et al. 1984). Our group has evaluated a number of behavioral attributes in patients with and without SMI. Only inwardly directed anger came out as a behaviour more likely to be found in those with SMI. Cigarette smoking has been demonstrated to decrease regional myocardial perfusion which is not associated with pain (Deanfield et al. 1986). Cold provocation has been shown to cause asymptomatic disturbances of regional myocardial perfusion (Shea et al. 1987). Perhaps there are other risk factors currently unknown, such as vulnerability to emotional stress, that are predictive for developing asymptomatic ischemia. A recent study by Freeman et al. (1987) demonstrated that psychological stress could significantly increase the frequency of asymptomatic ischemia. Further investigation is necessary to explore other existent risk factors.

It is now known that when a coronary artery is acutely occluded a temporal sequence is initiated, beginning with a metabolic abnormality (e.g., lactate production) followed by a hemodynamic abnormality (e.g., elevated left ventricular end-diastolic pressure, LVEDP, or an abnormality in myocardial relaxation) thence a mechanical abnormality (e.g., segmental wall motion defect), followed by an electrocardiographic change (e.g., ST segment depression). Finally, at the end of this temporal sequence, pain occurs. In one clinical

Table 3. Time course of ischemic echocardiography – PTCA: Onset after balloon inflation (seconds)

Vessel	Asynergy	ST shift	Angina
LAD	10.2	20.8	29
RCA	11.5	25.5	27.5

PTCA, percutaneous transluminal coronary angioplasty

study (Vissar 1986) echocardiography and continuous ECG monitoring was performed during percutaneous transluminal coronary angioplasty (PTCA) representing a model of acute coronary occlusion. From Table 3, one can see the temporal sequence of ischemia clinically displayed. Further, if the angioplasty balloon was deflated before the chest pain occurred, the sequence was reversed and an episode of SMI was experimentally induced.

The temporal sequence of ischemia discussed above has also been used to explain why some episodes of myocardial ischemia are painful and some painless. That is, in some instances the threshold of pain is not reached or the extent of ischemia is not great enough to trigger painful sensations.

Prevalence of SMI

To date, no epidemiologic studies have been reported which reliably assess the prevalence/incidence or prognostic importance of asymptomatic ischemia in the general population, or the population of persons with known coronary heart disease. The estimated prevalence of asymptomatic ischemia in the USA population of patients with stable angina pectoris has been estimated by Pepine (1986) to be 3 Million, based on the observations that 75 % of patients with angina have been demonstrated to have asymptomatic ischemia, and the total population of angina pectoris patients in the United States of America is 4 Million.

As already mentioned, SMI is common, but for obvious reasons there are only a few studies that have assessed its occurrence in totally asymptomatic persons (Table 2). The most extensive study was performed in Norway (Erikssen et al. 1976). In the years 1972–1975, a group of 2014 men aged 40–90 years who worked in one of five companies and governmental offices in Norway and who were healthy and presumably normal were recruited. A comprehensive baseline cardiovascular examination including a near maximal bicycle exercise test was performed. Of this group, 115 (5.7 %) had fulfilled one or more coronary heart disease suggestive criteria, 85 had a positive exercise ECG response, and in 75 a positive ECG during and/or after exercise was present. Further, of the 115 persons with suggestive criteria, 105 had angiography and 690 had greater than 50 % stenosis in one or more coronary arteries. In the 69 with positive angiograms, 58 had a positive exercise ECG (sensitivity 85 %). Furthermore, there were approximately equal distributions of single, double, and triple vessel disease. Fifty of the 69 men were totally asymptomatic even during the exercise test and were followed for a mean of 8 years. During this follow-up, 21 (42 %) suffered symptomatic coronary heart disease (18 had an MI or developed angina), while 3 died (2 suddenly).

In the United States Air Force (USAF) study of 1390 asymptomatic crewmen (mostly between ages 35–55) screened for latent CAD by exercise testing, 111 had positive exercise tests and 34 had coronary disease when angiography was performed (Froelicher et al. 1974). Eighteen of these had multivessel disease. In a Yale University study, 16 (of 129 men screened) had positive exercise tests, and 12 of these had angiographic evidence of coronary disease. In a 3 year follow-up, four had some cardiac event. Finally, Borer et al. (1975) performed exercise tests (bicycle ergometry, body position not specified) and coronary angiography in 89 patients with type II hyperlipoproteinemia. Of the total group, 30 were asymptomatic (26 men and 4 women). Eleven patients (37 %) had more than 50 % stenosis in at least one major coronary artery, while 19 (63 %) had either totally normal coronaries (10 subjects) or only mild irregularities (9 patients). Thus, the predictive value of a positive ST response for coronary disease in screening asymptomatic men was 66 %, 31 %, 75 %, and 37 %, respectively, in those four studies (i.e., the false-positive rates were 34 %, 69 %, 25 %, and 63 %). The broad predictive range undoubtedly relates to a number of factors, but subject selection is perhaps one of the most important. Erikssen et al. (1976) evaluated middle-aged Norwegian workmen, some of whom, despite being "asymptomatic", answered positive to a WHO questionnaire or had angina during their exercise test. The USAF study included subjects who were somewhat younger and who had testing performed as part of their evaluation to maintain flying status. In Langou et al.'s study (1980), 129 middle-aged men volunteered for a prospective study which included exercise testing and cardiac fluoroscopy. A questionnaire supplemented by direct questioning excluded any history of angina pectoris, myocardial infarction, or congestive heart failure. In Borer et al.'s report (1975), a subgroup of 30 hyperlipoproteinemic subjects were evaluated. It should also be noted that the type of exercise test varied (bicycle exercise, treadmill exercise, Masters test were also utilized) and exercise protocol lead systems were different.

Prognosis of SMI

There is mounting evidence that SMI adversely effects the prognosis of patients with CAD even when other baseline characteristics are considered. This has been evaluated in patients with chronic stable angina, unstable angina, asymptomatic patients with CAD, and in patients after myocardial infarction. For instance, a number of studies have demonstrated that the 1 year mortality for patients who have survived myocardial infarction and have been discharged from the hospital is 9 %. Within this 9 %, some individuals have a risk as low as 2 %, while others may have a risk as high as 25 %. In recent years, the exercise test usually utilizing a lower level of work compared with the standard test has been utilized to help stratify the risk of the post myocardial infarction patient. A number of observations have been made regarding the use of low level exercise testing after myocardial infarction. Included in these observations is the fact that 20%–25% of patients so treated have asymptomatic ST segment depression, and that risk is predicted much more by this than by the occurrence of exercise test induced angina. In the study by Theroux et al., the 1 year mortality was 2.1% in patients without ST changes during exercise and 27% in those with depression of the ST segment. Sudden death occurred in 0.7% without and 16% with ST segment depression. Exercise induced angina only predicted the subsequent occurrence of angina, but not mortality (when angina and ST depression occurred, 1 year mortality was 27%, when ST depression occurred alone, 1 year mortality was 26%).

Treatment of SMI

In the past medical and surgical treatment has been aimed at reducing the number of anginal episodes. Given the previous discussion the question now to be answered is whether we need to treat asymptomatic episodes as well. Critical to this question is whether therapy will alter the adverse outcome due to SMI. This question will not be answered for perhaps 5–10 years (studies are just now beginning to address this). In the meantime it would seem prudent to consider antiischemic therapy, in the belief that it will alter outcome.

Baseline therapy in all patients with CAD should, of course, consist of risk factor modification. As to the effect of treatment selected from the three major classes of antianginal agents (nitrates, beta adrenergic blocking agents, and calcium antagonists) and from within each class (e.g., nifedipine or diltiazem) controversy still exists. In 1983, our group began to evaluate patients with chronic stable angina pectoris for the prevalence of AECG SMI and its response to therapy aimed at alleviating the symptomatic episodes (Tables 4, 5). Of the first 72 patients, SMI was present in 64%. There was an average of 2.3 episodes of SMI/24 h of AECG. With any form of monotherapy there was a mean decrease in the frequency of SMI by 53%, with any dual therapy com-

Table 4. Impact of monotherapy on asymptomatic ischemia[a]

Therapy	Total hours	Episodes		ST product	
		Total	24 h	Total	24 h
Ca^{++}	600	39	1.6	2446	98
Beta B	1512	73	1.2	2078	33
Nitrates	576	11	0.5	143	42*

Ca^{++}, calcium antagonists; Beta B, beta adrenergic
* $p < 0.01$; compared to other forms of monotherapy
[a] $n = 60$; total 2688 h; 46 h/patient

Table 5. Impact of dual therapy on asymptomatic ischemia[a]

Therapy	Total hours	Episodes		ST product	
		Total	24 h	Total	24 h
Ca^{++} + BB	408	5	0.3	2.6	2
BB + Nit.	864	27	0.8	1591.0	44

Abbreviations: Ca^{++}, calcium antagonists; Beta B, beta adrenergic blockers; Nit., nitrates
N.S. between therapies
[a] $n = 28$; total 1344 h; 48 h/patient

bination SMI was decreased by 80%. In contrast, some studies have demonstrated that nifedipine had a significant or no effect on the frequency of SMI. In a recent study of a head to head comparison of diltiazem, nifedipine, and propranolol, compared to placebo, the beta blocking agent had the greatest effect, diltiazem an intermediate, and nifedipine no effect on SMI frequency.

Studies following successful PTCA and coronary bypass surgery have predictably resulted in a decrease in SMI, but the overall benefit of such an approach in asymptomatic patients remains to be determined.

Addendum

Silent myocardial ischemia is frequent in all patients with coronary artery disease and has been proven to impact unfavorably in the prognosis of patients with chronic stable angina, unstable angina, post myocardial infarction and patients with coronary disease who are totally asymptomatic. It has also been linked to some cases of sudden cardiac death. Suppression of silent ischemic episodes has been demonstrated with treatment of risk factors, all classes of antianginal agents and following mechanical interventions, however further evaluation of the best approach requires more investigation. It is unknown how much suppression is necessary and even if treatment favorably reverses the adverse prognosis imparted by silent ischemia.

References

Borer JS, Brensike JF, Redwood DR et al. (1975) Limitations of the electrocardiographic response to exercise in predicting coronary-artery disease. N Engl J Med 293/8:367–371

Cohn PF (1985) Silent myocardial ischemia: Classification, prevalence, and prognosis. Am J Med 79/3A:2–6

Deanfield JE, Maseri A, Selwyn AP (1983) Myocardial ischemia during daily life in patients with stable angina: Its relation to symptoms and heart rate changes. Lancet I:753–758

Deanfield JE, Kensett M, Wilson RA, Shea M, Horlock P, de Landsheere CM, Selwyn AP (1984) Silent myocardial ischemia due to mental stress. Lancet II:1001–1004

Deanfield JE, Shea MJ, Wilson RA, Horlock P, de Landsheere CM, Selwyn AP (1986) Direct effects of smoking on the heart: Silent ischemia disturbances of coronary flow. Am J Cardiol 57:1005–1009

Erikssen J, Enge I, Forfang K, Storatein O (1976) False positive diagnostic tests and coronary angiography findings in 105 presumably healthy males. Circulation 54/3:371–376

Freeman LJ, Nixon PGF, Sallabank P, Reaveley D (1987) Psychological stress and silent myocardial ischemia. Am Heart J 114/3:477–482

Froelicher F Jr, Thomas MM, Pillow C et al. (1974) Epidemiologic study of asymptomatic men screened by maximal treadmill testing for latent coronary artery disease. Am J Cardiol 34:770

Glasser SP (1987) Asymptomatic myocardial ischemia. Curr Prospect Coronary Care 1:67–73

Gottlieb SO, Weisfeldt ML, Duyang P, Mellits ED, Gerstenblith G (1986) Silent ischemia as a marker for early unfavorable outcomes in patients with unstable angina. N Engl J Med 314:1214–1219

Imperi GA, Pepine DJ (1986) Silent myocardial ischemia during daily activities: Studies in asymptomatic patients and those with various forms of angina. Cardiol Clin 4:635–642

Kawanishi DT, Rahimtoola SH (1987) Silent myocardial ischemia. Curr Probl Cardiol 12/9:516–566

Koehn DK, Glasser SP (1989) The impact of antianginal drug therapy on asymptomatic myocardial ischemia. J Clin Pharmacol 29:722–727

Langou RA, Huang EK, Kelley MJ, Cohen LS (1980) Predictive accuracy of coronary artery calcification and abnormal exercise test for coronary artery disease in asymptomatic men. Circulation 62/6:1196–1203

Maseri A (1987) Angina pectoris: Expanded concepts. Adv Intern Med 32:27–68

Nabel EG, Rocco MB, Selwyn AP (1987) Characteristics and significance of ischemia detected by ambulatory electrocardiographic monitoring. Circulation 75/5s:74–88

Pepine CJ (1986) Silent myocardial ischemia: Definition, magnitude, and scope of the problem. Cardiol Clin 4:627–633

Rocco MB, Barry J, Campbell S, Nabel E, Cook EF, Goldman L, Selwyn AP (1987) Circadian variation of transient myocardial ischemia in patients with coronary artery disease. Circulation 75/2:395–400

Rocco MB, Nabel EG, Campbell S, Goldman L, Barry J, Mead K, Selwyn A (1988) Prognostic importance of myocardial ischemia detected by ambulatory monitoring in patients with stable coronary artery diseases. Circulation 78:877–884

Rozanski A, Berman DS (1987) Silent myocardial ischemia. I. Pathophysiology, frequency of occurrence, and approaches toward detection. Am Heart J 114/3:615–626

Schang SJ, Pepine CJ (1977) Transient asymptomatic S-T segment depression during daily activity. Am J Cardiol 39:396

Shea MJ, Deanfield JE, de Landsheere CM, Wilson RA, Kensett M, Selwyn AP (1987) Asymptomatic myocardial ischemia following cold provocation. Am Heart J 114/3: 469–476

Sokolow M, McIlroy MB (1986) Clinical cardiology. Appleton-Century-Crofts, New York

Specchia G, de Servi S, Falcone C, Gavazzi A, Angoli L, Bramucci E, Ardissino D, Mussini A (1984) Mental arithmetic stress testing in patients with coronary artery disease. Am Heart J 108/1:56–63

Stern S, Tzivoni D (1973) Early detection of silent ischaemic heart disease by 24-hour electrocardiographic monitoring of active subjects. Br Heart J 36:481–486

Stern S, Tzivoni D, Stern Z (1974) Diagnostic accuracy of ambulatory ECG monitoring in ischemic heart disease. Circulation 52

Theroux P, Waters DD, Halphen C et al. (1979) Prognostic value of exercise testing soon after myocardial infarction. N Engl J Med 301:341–345

Vissar CA, David GK, Kan G, Romijn KN, Meltzer RS, Koblen JM, Dunning AJ (1986) Two-dimensional echocardiography during PTCA. Am Heart J 111:1035–1041

Pharmakotherapie der symptomatischen und asymptomatischen Myokardischämie

J. Dirschinger und W. Rudolph

Der Einsatz von Nitraten, nitratähnlichen Substanzen, β-Rezeptorenblockern, Kalziumantagonisten und antithrombotisch wirksamen Pharmaka bei Myokardischämie ist auf ein breites Spektrum klinischer Erscheinungsformen gerichtet. Dieses Spektrum umfaßt symptomatische Patienten, die entweder Angina pectoris ohne oder mit ST-Streckensenkungen aufweisen, und asymptomatische Kranke, bei denen nur ST-Veränderungen vorliegen. Die Daten der CASS-Registratur belegen nicht nur, daß Patienten mit koronarer Herzerkrankung gegenüber solchen ohne Koronarerkrankung eine eingeschränkte Prognose aufweisen, sie zeigen auch, daß die Letalität bei Patienten mit Myokardischämie, unabhängig davon, ob eine symptomatische oder asymptomatische Ischämie vorliegt, höher liegt als bei solchen ohne Myokardischämie (Weiner et al. 1987). Im folgenden soll dargestellt werden, wie wirksam eine Myokardischämie beeinflußt werden kann und ob eine effektive antiischämische und antithrombotische Therapie mit einer Verbesserung der Prognose gleichzusetzen ist oder ob die Pharmakotherapie in erster Linie als symptomatische Maßnahme anzusehen ist.

Nitrate und Molsidomin

Drei Nitratderivate, Glyceroltrinitrat (GTN), Isosorbiddinitrat (ISDN) und Isosorbid-5-mononitrat (IS-5-MN), sowie die nitratähnliche Substanz Molsidomin finden gegenwärtig im wesentlichen klinische Anwendung. GTN wird überwiegend sublingual zur Akutbehandlung bzw. -prophylaxe des Angina-pectoris-Anfalls, neuerdings transdermal zur Langzeitprophylaxe eingesetzt. Bei ISDN, IS-5-MN und Molsidomin steht die orale Anwendung retardierter und nichtretardierter Präparate zur Langzeitbehandlung im Vordergrund.

Eine auf Dauer wirksame Behandlung ohne Toleranzentwicklung ist mit Nitraten nur als sog. Intervalltherapie möglich (Rudolph et al. 1983; Rudolph u. Blasini 1984). Die Effektivität des Konzepts der intermittierenden Wirkstoffverabreichung ist in mehreren Studien belegt: Sowohl bei Verabreichung von 20 mg ISDN am Morgen und am Mittag, also unter Einhaltung eines 19stündigen, einnahmefreien Intervalls, als auch bei einmaliger täglicher Gabe einer hohen Nitratdosis von 120 mg retardiertem ISDN bzw. 50 oder 100 mg retardiertem IS-5-MN, ebenso wie bei Verwendung von Nitroglycerinpflastern unter Inkorporation eines 12stündigen pflasterfreien Intervalls, findet sich

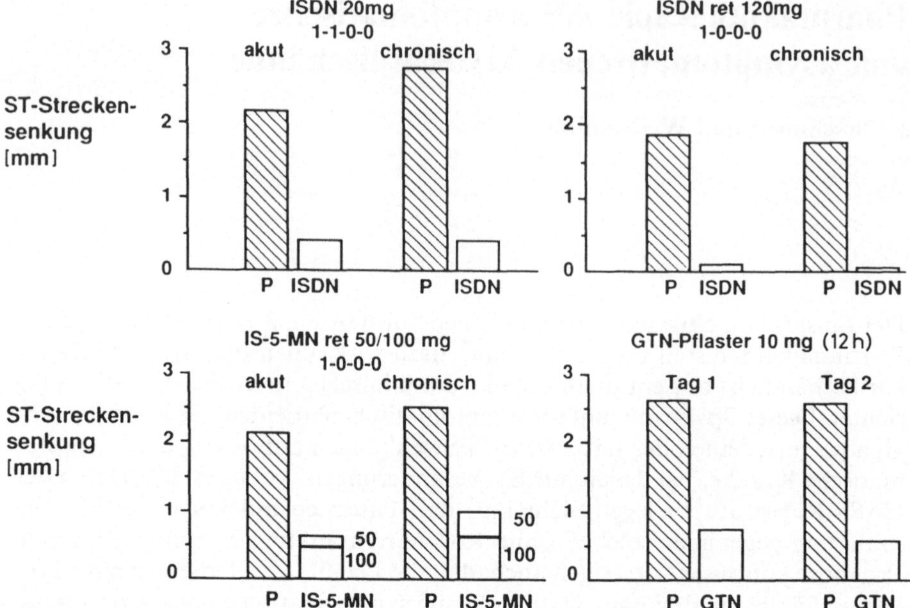

Abb. 1. Intervalltherapie mit verschiedenen Nitraten. Nach Akutverabreichung wie unter Langzeittherapie mit ISDN, ISDN retard und IS-5-MN, als auch bei wiederholter 12stündiger Applikation eines GTN-Pflasters zeigt sich ein jeweils vergleichbarer Rückgang der belastungsinduzierten ST-Streckensenkung (*ISDN* Isosorbiddinitrat, *ISDN ret* Isosorbiddinitrat retard, *IS-5-MN* Isosorbid-5-mononitrat, *GTN* Glyceroltrinitrat, *P* Placebo)

nach Erstverabreichung und auch unter Langzeittherapie ein vergleichbarer antiischämischer Effekt ohne Hinweis für eine Wirkungsabschwächung (Rudolph et al. 1988, 1989; Abb. 1).

Hingegen führen alle Therapieschemata, die auf weitgehend konstante Wirkstoffkonzentrationen abzielen, zur Toleranz, wie dies für die Verabreichung von ISDN retard in einer Dosierung von 3mal 20, 3mal 40 und 3mal 60 mg, für nichtretardiertes ISDN in einer Dosierung von 4mal 40 mg und für Nitroglycerinpflaster mit Freisetzungsraten zwischen 10 und 30 mg pro 24 h gezeigt wurde (Rudolph et al. 1981, 1983; Rudolph u. Blasini 1984; Rudolph et al. 1985; Reiniger u. Rudolph 1985; Reiniger et al. 1985, 1987; Abb. 2).

Toleranz bzw. erhaltene Wirksamkeit lassen sich aus dem Verhalten der Plasmaspiegel erklären. Im Stadium der Toleranz, hervorgerufen durch eine Medikation mit 4mal 40 mg ISDN täglich, finden sich hohe Plasmakonzentrationen, insbesondere von IS-5-MN. Diese steigen zwar nach erneuter Verabreichung von ISDN weiter an, die Änderung in Relation zum hohen Basalwert ist jedoch offenbar zu gering, um eine vaskuläre Antwort zu induzieren. Im Gegensatz dazu findet sich bei den dargestellten Schemata der Intervalltherapie keine wesentliche Kumulation der Nitrate im Plasma, so daß nach Akutgabe wie unter chronischer Medikation die jeweilige Medikamentengabe zu

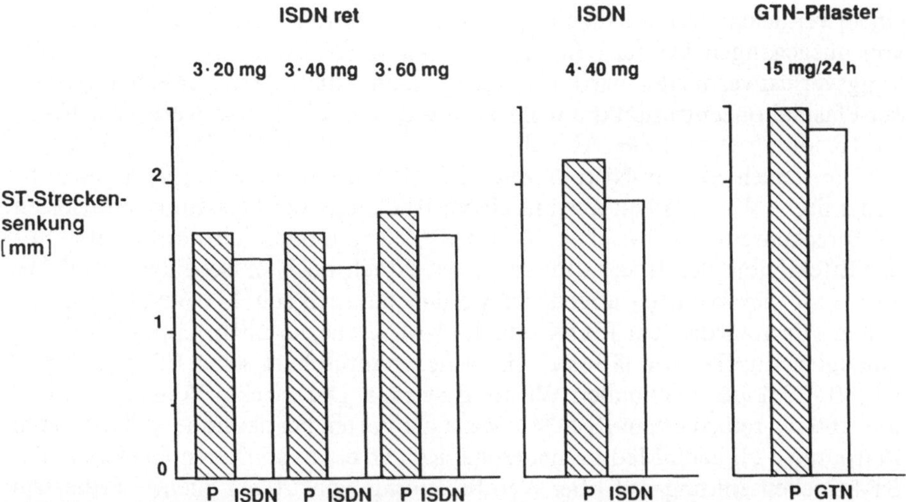

Abb. 2. Toleranzentwicklung unter verschiedenen Nitraten. Bei Dosierungsschemata, die auf einen 24 h lang anhaltenden Schutz abzielen, findet sich kein signifikanter Rückgang der belastungsinduzierten ST-Streckensenkung

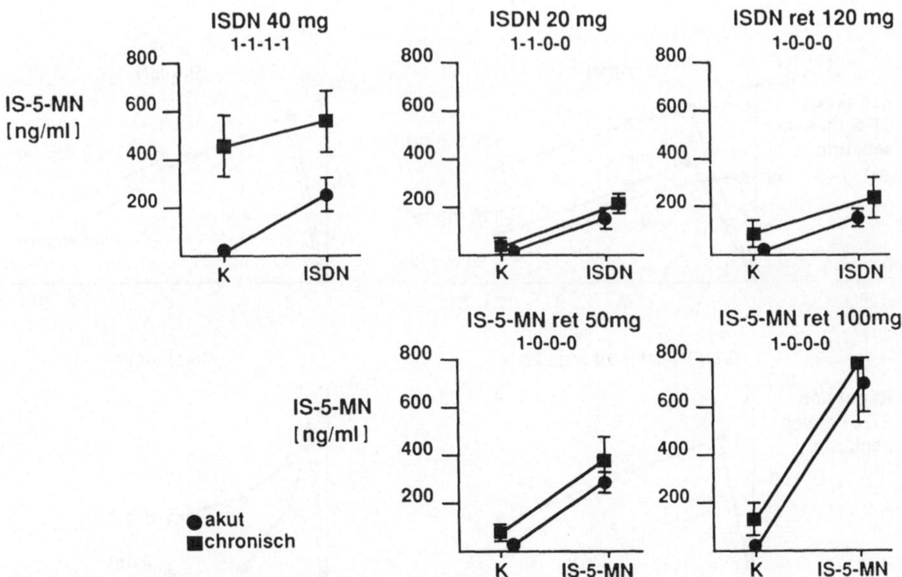

Abb. 3. Verhalten der IS-5-MN-Plasmakonzentrationen nach Akutverabreichung und unter Langzeittherapie bei Intervalltherapie und bei einer zur Toleranz führenden Medikation (ISDN 4mal 40 mg; *K* Kontrolle vor Substanzgabe)

einem vergleichbaren Anstieg der Plasmakonzentrationen führt. Es kann davon ausgegangen werden, daß eine Wirkungsabschwächung bzw. ein Wirkungsverlust vermieden wird, wenn die erneute Substanzzufuhr einen Anstieg der Plasmakonzentration um mehr als das 2,5fache des Basalwertes zur Folge hat (Rudolph et al. 1988; Abb. 3).

Die verschiedenen Nitrate zeichnen sich durch eine ausgeprägte antiischämische Wirksamkeit aus mit einem Rückgang der belastungsinduzierten ST-Streckensenkung zwischen 60 und 90 % 2–3 h nach Substanzzufuhr. Für die Effektivität der Intervalltherapie ist jedoch auch entscheidend, welcher Zeitraum therapeutisch abgedeckt werden kann. Noch 12 h nach Verabreichung von retardiertem ISDN und IS-5-MN, ebenso nach Applikation von Nitroglycerinpflastern, läßt sich ein sicherer antiischämischer Effekt belegen, der 40–50% der maximalen Wirkung beträgt. Der direkte Vergleich von 50 und 100 mg retardiertem IS-5-MN weist als Vorteil bei der höheren Dosis zum Zeitpunkt 12 h nach Medikamentengabe einen etwas stärkeren Rückgang der ST-Streckensenkung auf. Bei Verabreichung von 20 mg nichtretardiertem ISDN am Morgen und am Mittag ist ein antiischämischer Schutz nur über einen kürzeren Zeitraum gewährleistet. 12 h nach der ersten Medikamentengabe ist keine Beeinflussung der ST-Streckensenkung mehr nachweisbar (Rudolph et al. 1989; Abb. 4).

Die Intervalltherapie bietet somit keinen 24 h anhaltenden Schutz. Ischämische Episoden treten jedoch überwiegend während des Tages, mit Häufigkeitsgipfeln am Morgen und am Nachmittag, auf. Diese werden z. B. durch 120 mg retardiertes ISDN wirksam beeinflußt. Die therapeutische Lücke

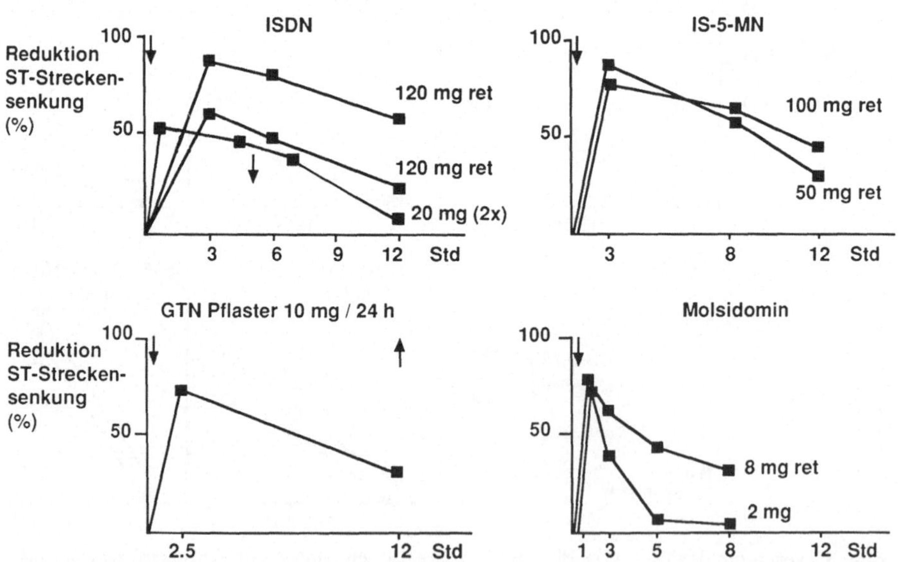

Abb. 4. Wirkungsausmaß und Wirkungsdauer von Nitraten und Molsidomin in unterschiedlicher Dosierung

bleibt für den Zeitraum bestehen, in dem es ohnehin selten zu ischämischen Episoden kommt. In der Phase der nachlassenden Nitratwirkung sind auch keine Reboundphänomene, weder bei einer Intervalltherapie mit oral verabreichten Nitraten noch bei Verwendung von GTN, zu befürchten. So war nach Beendigung einer GTN-Medikation weder während des ST-Streckenmonitorings eine Häufung ischämischer Episoden noch bei Belastungsuntersuchungen eine Änderung im ST-Streckenverhalten, der Herzfrequenz oder des Blutdrucks zu verzeichnen (Rudolph et al. 1989).

Insgesamt kann aufgrund eigener Befunde davon ausgegangen werden, daß eine Intervalltherapie mit Nitraten zu einer Reduktion der Angina-pectoris-Anfallshäufigkeit um etwa 50 % führt. Der Schweregrad der Angina pectoris, beurteilt nach der Klassifikation der Canadian Cardiovascular Society, geht ebenfalls um etwa 50 % zurück, und bei 20–30 % der Patienten wird Beschwerdefreiheit erreicht. Langzeit-EKG-Registrierungen weisen einen Rückgang ischämischer Episoden – symptomatische und asymptomatische – um etwa 50 % auf.

Nicht gelöst mit der Intervalltherapie ist das Problem der frühen Toleranzentwicklung, das sowohl bei oraler Medikation als auch, in besonderem Maße, beim Einsatz von Nitroglycerinpflastern offenkundig ist. So schwächt sich bei gleichbleibend hohen Plasmakonzentrationen der 2½ h nach Applikation eines transdermalen GTN-Systems ausgeprägte antiischämische Effekt innerhalb von 8–12 h deutlich ab (Reiniger u. Rudolph 1985). Diese frühe Toleranzentwicklung läßt sich vermeiden, wenn die Plasmanitroglycerinkonzentrationen über 12 h ständig ansteigen. Durch Applikation von 4 Nitroglycerinpflastern mit einer jeweiligen Freisetzungsrate von 5 mg werden derartig kontinuierlich ansteigende Plasmakonzentrationen und damit ein konstanter antiischämischer Effekt über nahezu 12 h erreicht. Das anschließende 12stündige pflasterfreie Intervall ist notwendig, um ein Wiederansprechen am nächsten Morgen zu garantieren (Rudolph et al. 1989). Transdermale Systeme, aber auch orale Präparate, sollten eine derartige Freisetzungskinetik aufweisen, um über 12 h einen konstant ausgeprägten antiischämischen Effekt zu gewährleisten.

Pharmakologische Ansätze, eine Nitrattoleranz zu verhindern, stellen die zusätzliche Verabreichung von SH-Donatoren dar. Klinische Untersuchungen bei Patienten mit Herzinsuffizienz bzw. koronarer Herzerkrankung, die über eine Aufhebung der Nitrattoleranz durch Acetylcystein berichten, stehen Mitteilungen gegenüber, daß eine Nitrattoleranz nicht durchbrochen werden konnte bzw. das Auftreten einer Toleranz nicht verhindert wurde. Die unterschiedlichen Resultate könnten sowohl in der Art des verwandten Nitrats – ISDN oder GTN – als auch in unterschiedlich verabreichten Dosen von N-Acetylcystein liegen. Offenbar sind große Substanzmengen notwendig, um einen Effekt zu erzielen. Von Interesse scheint darüber hinaus die Beobachtung zu sein, daß nur das L-Isomer des Acetylcysteins, nicht jedoch das D-Isomer, geeignet ist, eine Toleranz zu verhindern (Packer et al. 1987; Parker et al. 1987; Hogan et al. 1989). Die Erfahrungen mit anderen SH-Donatoren, wie Captopril und Methionin, sind begrenzt. In eigenen Untersuchungen zeigte sich bei

Patienten, die mit einem 20-mg-GTN-Pflaster vorbehandelt waren, keine Beeinflussung der belastungsinduzierten ST-Streckensenkung durch Captopril. Allerdings lag bei diesen Patienten lediglich eine Wirkungsabschwächung und nicht das Vollbild einer Nitrattoleranz vor, so daß möglicherweise ein wenig ausgeprägter Captoprileffekt nicht sichtbar wurde. Andere, kürzlich mitgeteilte Befunde ergaben, daß Captopril einer nitratbedingten Wirkungsabschwächung mäßig entgegenwirkt (de Graef et al. 1989).

Die pharmakologischen Wege, eine Nitrattoleranz zu verhindern, befinden sich somit nach wie vor im klinischen Erprobungsstadium. Die einzige Möglichkeit, Nitrate effektiv in der Langzeittherapie einzusetzen, stellt nach wie vor die Intervalltherapie dar.

Molsidomin hat sich, den organischen Nitraten vergleichbar, ebenfalls als ausgeprägt antiischämisch wirksam erwiesen. In einer Vergleichsuntersuchung zeigte sich 1 h nach Verabreichung von 2 mg nichtretardiertem als auch von 8 mg retardiertem Molsidomin ein Rückgang der belastungsinduzierten ST-Streckensenkung um 74 %. Als nicht ausreichend lange wirksam ist das Standardpräparat anzusehen. Bereits 5 h nach Medikation war keine Beeinflussung der ST-Streckensenkung mehr zu belegen. Für 8 mg retardiertes Molsidomin zeigte sich hingegen über 8 h ein im zeitlichen Verlauf sich abschwächender antiischämischer Effekt. Zu diesem Zeitpunkt betrug der Rückgang der belastungsinduzierten ST-Streckensenkungen noch 40 % des maximalen Effekts (Rudolph u. Dirschinger 1985).

Damit ergibt sich für die Langzeittherapie in der Regel eine 3- bis 4malige tägliche Verabreichung retardierten Molsidomins. Diese mehrfache tägliche Substanzzufuhr hat im Gegensatz zu derjenigen von Nitraten keine Toleranzentwicklung hinsichtlich der antianginösen und antiischämischen Wirkung zur Folge (Beyerle u. Rudolph 1986). Die fehlende Toleranzentwicklung unter Molsidomin wird darauf zurückgeführt, daß thiolunabhängig die Guanylatcyclase stimuliert wird. Für diese Annahme spricht die eigene Beobachtung, daß Molsidomin auch im Stadium der Nitrattoleranz noch wirksam ist. Wird durch Nitroglycerinpflaster eine Toleranz induziert, kenntlich am Rückgang der Beeinflussung der ST-Streckensenkung, zeigt sich bei Verabreichung eines erneuten Nitroglycerinpflasters zusammen mit Molsidomin wieder eine deutliche Reduktion der belastungsinduzierten ST-Streckensenkung, während die alleinige Nitratzufuhr keinen signifikanten Effekt zur Folge hat.

Keine bzw. keine ausreichenden Daten liegen zur Beantwortung der Frage vor, ob durch Nitrate oder Molsidomin der koronarsklerotische Prozeß oder die Prognose der koronaren Herzerkrankung beeinflußt wird. Retrospektive Analysen, die für Patienten nach überstandenem Myokardinfarkt unter Nitratbehandlung höhere Überlebensraten zeigen, weisen erhebliche methodische Einschränkungen auf, die es nicht zulassen, einen prognostisch günstigen Effekt von Nitraten abzuleiten.

β-Rezeptorenblocker

Die zahlreichen verschiedenen β-Rezeptorenblocker lassen sich hinsichtlich ihrer β_1-Selektivität, des Vorhandenseins einer intrinsischen sympathikomimetischen Eigenaktivität (ISA), ihrer Hydrophilie bzw. Lipophilie und des Vorhandenseins vasodilatierender Eigenschaften klassifizieren. Die Wahl des β-Rezeptorenblockers wird durch diese pharmakologischen Eigenschaften bestimmt. Eine β_1-Selektivität verspricht eine höhere Verträglichkeit, da spezifische unerwünschte Wirkungen, z. B. auf Bronchien, Gefäße, endokrine Organe, Leber und Muskeln, überwiegend durch β_2-Rezeptoren vermittelt werden. Für die Verwendung von Substanzen ohne ISA spricht, daß für β-Blocker mit ISA kein sekundär präventiver Effekt gesichert wurde und daß bei Anwendung von β-Blockern mit ISA Nebenwirkungen wie Schlafstörungen, Alpträume, Unruhe und Tremor auftreten können, die auf die ISA bezogen werden. Hydrophile Antagonisten sind lange wirksam, was eine Einmaldosierung ermöglicht. Bei Verwendung lipophiler Substanzen setzt dies eine entsprechende Retardierung voraus. Darüber hinaus scheinen zentralnervöse Effekte und Medikamenteninteraktionen unter hydrophilen β-Rezeptorenblockern geringer ausgeprägt zu sein. Mit einer neuen Generation von β-Rezeptorenblockern, die zusätzlich vasodilatierende Eigenschaften – α_1- oder β_2-rezeptorenvermittelt und/oder α-/β-rezeptorenunabhängig – aufweist, wird eine Senkung des peripheren Widerstands, eine geringere Kardiodepression und eine verminderte Wirkung auf die Bronchien und damit eine höhere Effektivität und eine bessere Verträglichkeit angestrebt (Borchard 1988).

Bei der Vielzahl verschiedener β-Rezeptorenblocker wird im folgenden auf Bisoprolol, Atenolol und Metoprolol näher eingegangen, weil diese Substanzen β_1-selektiv sind, weil sie keine ISA aufweisen und weil die Hydrophilie von Bisoprolol und Atenolol bzw. die Verfügbarkeit eines Retardpräparates des lipophilen Metoprolol eine lange Wirkungsdauer versprechen.

So findet sich 3 h nach Gabe von 5–10 mg Bisoprolol, 50–100 mg Atenolol oder 200 mg Metoprolol in Retardform eine Zunahme der Belastungskapazität zwischen 30 und 50 % und ein Rückgang der belastungsinduzierten ST-Streckensenkung zwischen 60 und 70 %. 24 h nach Verabreichung von Atenolol und Bisprolol beträgt die antiischämische Wirkung noch 50–60 % des maximalen Effekts. Der Vorteil höherer Dosierungen von 10 mg Bisoprolol bzw. 100 mg Atenolol liegt in einer stärkeren antiischämischen Wirkung weniger 3 h als vielmehr 24 h nach Medikamenteneinnahme. Weitere Dosissteigerungen erbringen in der Regel weder hinsichtlich Wirkungsausmaß noch Wirkungsdauer einen Zugewinn (Jackson et al. 1980; de Muinck et al. 1987; Tabelle 1).

Auch für retardiertes Metoprolol in einer Dosierung von 200 mg wurde eine über 24 h nachweisbare Steigerung der Belastungskapazität gezeigt. (Uusitalo u. Keyrilainen 1979). In eigenen Untersuchungen konnte allerdings 24 h nach Medikamentengabe auf vergleichbarer Belastungsstufe kein Rückgang der belastungsinduzierten ST-Streckensenkung gesichert werden (Lehmann et al. 1988). Insgesamt ist davon auszugehen, daß durch diese β-Rezeptoren-

Tabelle 1. Wirkungsausmaß und Wirkungsdauer verschiedener β-Rezeptorenblocker (Angaben in %) [a] Vergleichbare Belastungsstufe; [b] Maximale Belastung

Autoren	Substanz/Dosis		Zunahme Belastungskapazität		Rückgang ST-Streckensenkung	
			2/3 h	24 h	3 h	24 h
de Muinck et al. 1987	Bisoprolol	5 mg	44	54	62	26[a]
		10 mg	17	35	66	40[a]
Jackson et al. 1980	Atenolol	50 mg	34	17	34	12[b]
		100 mg	30	20	50	47[b]
Uusitalo u. Keyriläinen 1979	Metoprolol duriles	200 mg	29	15		
Lehmann et al. 1988					73	0[a]

blocker eine Reduktion der Angina-pectoris-Anfallshäufigkeit um 40–60 % erreicht wird und daß die Häufigkeit ischämischer Episoden im Langzeit-EKG um 60–70 % abnimmt. Dabei lassen sich offenbar auch ischämische Episoden, die nicht aus einer Erhöhung des O_2-Bedarfs resultieren, durch β-Rezeptorenblocker effektiv beeinflussen (Rudolph u. Dirschinger 1984; Chierchia et al. 1987; Imperi et al. 1987). Labetalol und Celiprolol, die bereits im Handel sind, sowie Carvedilol, Bucindolol, Dilevalol oder Medroxalol sind β-Rezeptorenblocker mit vasodilatierender Wirkung. Gemessen an der Beeinflussung der ST-Streckensenkung und der Belastungskapazität erwies sich Carvedilol in einer Dosierung von 25 mg nicht als ausreichend wirksam. Nach Verabreichung von 50 mg war ein deutlicher, mit der Gabe von 50 mg Atenolol vergleichbarer Effekt zu erzielen. Danach schlägt sich die vasodilatierende Eigenschaft nicht in einem stärkeren antiischämischen Effekt nieder. Ob sich aus dem Nachweis einer Zunahme der Ruheauswurffraktion, nicht der Belastungsauswurffraktion, eine bevorzugte Anwendung bei Patienten mit deutlich eingeschränkter linksventrikulärer Funktion ergibt, muß ebenso in weiteren klinischen Untersuchungen überprüft werden wie die Frage, ob und welche Vorteile die vasodilatierende Wirkung im Hinblick auf eine bessere Verträglichkeit bringt (Prichard u. Tomlinson 1989; Lahiri et al. 1989; Freedman et al. 1987).

β-Rezeptorenblocker sind in die Gruppe gut verträglicher Medikamente einzustufen. Die Häufigkeit unerwünschter Wirkungen wird bei Verwendung β_1-selektiver Substanzen mit etwa 4 %, nichtselektiver Substanzen mit etwa 10 % angegeben. Unerwünschte Wirkungen resulierten häufig aus den β-rezeptorenblockierenden Effekten selbst (Borchard 1988). Sie sind in der Regel deshalb nur zu erwarten, wenn der Einsatz dieser Substanzen bedingt oder nicht indiziert ist. Die Häufigkeitsangaben über das Auftreten einer sexuellen Dysfunktion unter Behandlung mit einem β-Rezeptorenblocker schwanken beträchtlich und beziehen sich fast ausschließlich auf unkontrollierte Beobachtungen an kleinen mit Propranolol behandelten Patientenkollektiven. Lediglich die MCR-Studie bei Patienten mit leichter bis mäßiger Hypertonie scheint gegenwärtig geeignet, das Problem der propranololassoziierten sexuellen Dysfunktion in die richtige Dimension zu bringen. Bei einer Impotenzrate in der Placebogruppe von 9–10% und in der β-Blockergruppe von 13–14% sind somit weniger als 5% der mit Propranolol behandelten Patienten betroffen. Es scheint gesichert, daß eine Dosisabhängigkeit vorliegt und bei Dosisreduktion bzw. Absetzen des β-Rezeptorenblockers Reversibilität besteht. Klinischen Erfahrungen zufolge – wenngleich unkontrolliert – läßt sich die Induktion einer sexuellen Dysfunktion auch bei Verwendung β_1-selektiver Substanzen nicht ausschließen, obwohl Untersuchungen vorliegen, die über gleiche Häufigkeitsraten unter Atenolol bzw. Bisoprolol und unter Placebogabe berichten. Insgesamt sprechen die vorliegenden Daten dafür, daß eine sexuelle Dysfunktion erheblich seltener durch β_1-selektive Substanzen als durch Propranolol ausgelöst wird und diese Nebenwirkung damit bei der überwiegenden Anzahl der Patienten keinen therapiebegrenzenden Faktor darstellt (McDevitt 1988; Bühler et al. 1986; Smith u. Talbert 1986).

Bezüglich der Beeinflussung des Lipidstoffwechsels zeigt sich, daß β_1-selektive Substanzen zu einer weniger ausgeprägten Zunahme der Triglyceridspiegel als Propranolol führen. Auch die unter β-Rezeptorenblockern zu verzeichnende Abnahme der HDL und die daraus resultierende Zunahme des LDL/HDL-Quotienten scheint unter selektiven Substanzen geringer ausgeprägt zu sein als unter der nichtselektiven Substanz Propranolol. Ob medikamentös induzierten Veränderungen der Blutlipide dieselbe Bedeutung für den Krankheitsverlauf zukommt wie spontan vorhandenen, ist gegenwärtig ungeklärt. Die Ergebnisse der Sekundärpräventionsstudien sprechen zumindest dafür, daß die prognostisch günstigen Effekte gegenüber möglichen ungünstigen Auswirkungen überwiegen (Miller 1987; Olsson u. Rehnqvist 1987; Herrmann et al. 1988; Roberts 1987; Frithz u. Weiner 1986).

Die Frage, ob β-Rezeptorenblocker einen Einfluß auf den koronarsklerotischen Prozeß haben, läßt sich gegenwärtig nicht beantworten. In einer klinischen Studie zeigte sich eine angiographisch nachweisbare Progression vorbestehender Läsionen bei 53% der Patienten, die Propranolol, und bei 47% der Patienten, die Nitrate erhielten. Diese geringen Unterschiede wurden als Hinweis auf eine möglicherweise durch β-Rezeptorenblocker verstärkte Progression der Erkrankung interpretiert.

Demgegenüber stehen Untersuchungen, die einen antiatherogenen Effekt von β_1-selektiven Substanzen wie Metoprolol belegen. Welche klinische Relevanz diesen unterschiedlichen Befunden zukommt, läßt sich nur durch entsprechend konzipierte Progressionsstudien klären (Loaldi et al. 1989; Ablad et al. 1988).

Welchen Einfluß β-Rezeptorenblocker auf Morbidität und Letalität ausüben, ist nur für Patienten nach überstandenem Myokardinfarkt zu beantworten. Aus den gepoolten Daten der zahlreichen Sekundärpräventionsstudien ist eine Reduktion des plötzlichen Herztodes abzuleiten, die sich letztlich auch in einer Reduktion der Gesamtsterblichkeit niederschlägt. Obwohl diese Reduktion in den Einzelstudien nichtsignifikant ist, ergeben diese gepoolten Daten insgesamt auch eine Reduktion der Reinfarktrate. Dieser prognostisch günstige Effekt, insbesondere die Reduktion der Reinfarktrate, läßt sich durch die antiischämische Wirkung allein nicht hinreichend erklären. Basierend auf der Beobachtung einer Korrelation zwischen Ausmaß der Herzfrequenzverlangsamung und Reduktion der Sterblichkeit und Infarktrate wird vielmehr angenommen, daß durch die Herzfrequenzverlangsamung die Spannungsbelastung eines arteriosklerotischen Plaques abnimmt und sich damit das Risiko einer Plaqueruptur mit nachfolgender Thrombose reduziert oder das Ereignis zumindest zeitlich nach hinten verschoben wird. Dieser Mechanismus sollte auch bei Koronarkranken ohne abgelaufenen Infarkt zum Tragen kommen (Fitzgerald 1987).

Kalziumantagonisten

Als Kalziumantagonisten werden derzeit überwiegend Nifedipin, Diltiazem und Verapamil eingesetzt. Daneben sind eine Reihe von Substanzen der 2. Generation in klinischer Prüfung, und einige davon bereits im Handel. Diese Neuentwicklungen, insbesondere aus der Dihydropyridinreihe, sind in erster Linie auf eine ausgeprägte Wirksamkeit bei hoher Vasoselektivität und auf eine lange Wirkungsdauer gerichtet. Die Übersicht bringt eine Auflistung der bis jetzt entwickelten Kalziumantagonisten:

1. Generation	Verapamil	Nifedipin		Diltiazem
2. Generation	Gallopamil	Nitrendipin	Nilvadipin	Bepridil
	Anipamil	Nisoldipin	Amlodipin	
	Tiapamil	Nicardipin	Isradipin	
		Nimodipin	Felodipin	

Gemessen am Rückgang der belastungsinduzierten ST-Streckensenkung um 60–70 % zeichnen sich Nifedipin, Verapamil und Diltiazem durch eine sichere antiischämische Wirkung aus. Jedoch auch bei Verabreichung hoher Dosen wie 20 mg retardiertem Nifedipin, 120 mg Diltiazem oder 120 mg Verapamil ist die Wirkungsdauer begrenzt. Nach 5–7 h beträgt der Rückgang der ST-Streckensenkung nur noch etwa 50 % des maximalen Effekts, so daß in der Regel eine 3- (bis 4)mal tägliche Applikation notwendig ist (Abb. 5).

Die Häufigkeit von Angina pectoris oder ischämischen Episoden im Langzeit-EKG geht unter diesen Substanzen um jeweils etwa 50 % zurück (Brügmann et al. 1983, 1984; Lehmann et al. 1989; Lynch et al. 1980; Findlay et al. 1986).

In eigenen Untersuchungen zeigte sich bei 2maliger Gabe von 10 mg Nisoldipin bzw. einmaliger Verabreichung von 10 mg Amlodipin ein über 24 h nahezu gleichbleibender antiischämischer Effekt mit einem Rückgang der ST-Streckensenkung um etwa 30 % (Brügmann et al. 1984). Damit liegt das Wirkungsausmaß unter dem, das wir unter den „klassischen“ Kalziumantagoni-

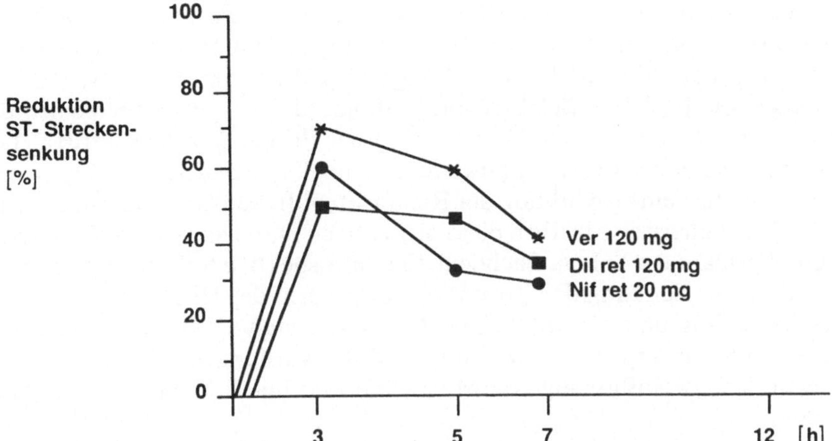

Abb. 5. Wirkungsausmaß und Wirkungsdauer von Nifedipin retard (*Nif ret*), Diltiazem retard (*Dil ret*) und Verapamil (*Ver*)

sten beobachteten. Möglicherweise lassen sich durch höhere Dosierungen, wie sie sich v.a. für Amlodipin abzeichnen, stärkere Effekte erzielen.

Eine Wertung der verschiedenen Substanzen der 2. Generation ist zum gegenwärtigen Zeitpunkt nur begrenzt möglich. Nach den vorliegenden Daten scheinen sie bezüglich antianginöser und antiischämischer Wirksamkeit nicht effektiver als die von Nifedipin, Diltiazem und Verapamil zu sein. Nicht nur Bepridil, dessen Einsatz durch die mögliche Induktion von Rhythmusstörungen begrenzt scheint, sondern auch die Mehrzahl der neueren Dihydropyridinderivate weisen im Gegensatz zu Nifedipin eine längere Wirkungsdauer auf, die eine nur 1- bis 2malige tägliche Medikamenteneinnahme ermöglicht. Ob die Vasoselektivität und die damit geringer ausgeprägte negativ inotrope Wirkung der Dihydropyridinderivate einen breiteren Einsatz bei Patienten mit erheblich eingeschränkter linksventrikulärer Funktion zuläßt, wird in Frage gestellt bzw. muß zumindest derzeit noch offenbleiben (Packer 1989a).

Experimentell belegte antiatherogene Effekte von Kalziumantagonisten bilden die Grundlage angiographischer Verlaufsuntersuchungen. In der INTACT-Studie zeigten sich unter Placebo bei 49% der Patienten, nach 3jähriger Behandlung mit 4mal 20 mg Nifedipin hingegen bei 41% der Patienten, neu aufgetretene koronarsklerotische Läsionen. Dies entspricht einer Reduktion um 26%. Die Progression vorbestehender Veränderungen wurde nicht beeinflußt (Lichtlen et al. 1989). Untersuchungen, die bereits nach 1jähriger Behandlung mit Nifedipin in erheblich geringerem Umfang eine Progression koronarsklerotischer Veränderungen erkennen ließen als in einem mit Nitraten behandelten Kollektiv, beziehen sich auf kleine Patientengruppen (Loaldi et al. 1989). Die Ergebnisse größerer, noch laufender Studien mit Nisoldipin und mit Verapamil (FIPS) müssen noch abgewartet werden. Die derzeit vorliegenden wenigen Daten sprechen eher für eine Verlangsamung des arteriosklerotischen Prozesses durch Kalziumantagonisten als dagegen.

Wie für β-Rezeptorenblocker liegen auch für die Kalziumantagonisten Nifedipin, Diltiazem und Verapamil zur Frage der Beeinflussung von Morbidität und Letalität nur Untersuchungen bei Patienten nach überstandenem Infarkt vor. In diesen Sekundärpräventionsstudien konnte für keinen dieser Kalziumantagonisten eine Verbesserung der Prognose belegt werden. Lediglich in der „Non-Q-wave-Infarktstudie" mit Diltiazem zeigte sich innerhalb von 2 Wochen eine Reduktion der Reinfarkthäufigkeit bei unveränderter Letalität. Möglicherweise bleiben diese protektiven Diltiazemeffekte über eine längere Zeitspanne erhalten, nachdem eine retrospektive Subgruppenanalyse zumindest eine Senkung der Koronarinzidenz, d. h. der Häufigkeit aller kardialen Todesfälle und nichttödlichen Reinfarkte, ergab. Mit Ausnahme der genannten Subgruppen ist somit aufgrund der vorliegenden Daten wohl keine wesentliche Beeinflussung von Morbidität und Letalität zu erwarten (Gibson 1989).

Kombinationstherapie

Die unterschiedlichen Wirkungsmechanismen der verschiedenen Substanzen lassen ihre Verabreichung in Kombination sinnvoll erscheinen. Dennoch wird der Wert einer Kombinationstherapie teilweise kontrovers diskutiert (Packer 1989 b). Einerseits führt man an, daß in einigen Studien, die die Überlegenheit einer Kombination zeigten, nicht geprüft wurde, ob mit einer durch Dosistitration ermittelten, optimalen Monotherapie derselbe Effekt erreichbar gewesen wäre. Andererseits erfolgt der Hinweis, daß Studien, die keine Überlegenheit belegen, vielfach bei Patienten durchgeführt wurden, bei denen bereits mit der Monotherapie der maximal mögliche Effekt nahezu erreicht wurde, so daß der Wirkungszuwachs einer Kombinationsbehandlung zwar noch erkennbar, statistisch aber nicht mehr zu sichern war. Wesentlich für die Bewertung scheint, welchem Endpunkt, Zunahme der Belastbarkeit oder Rückgang des Ischämieausmaßes, Bedeutung beigemessen wird. Untersuchungen, in denen beide Größen beurteilt wurden, verdeutlichen dies: So zeigte sich unter Diltiazem und Propranolol keine stärkere Zunahme der Belastungskapazität als unter alleiniger Diltiazemtherapie. Jedoch war unter der Kombinationsbehandlung ein deutlich ausgeprägterer Rückgang der belastungsinduzierten ST-Streckensenkung, also ein stärkerer antiischämischer Effekt, zu verzeichnen (Hung et al. 1983).

Eigene Daten weisen bei Kombination eines β-Rezeptorenblockers mit einem Nitrat oder bei Kombination eines β-Rezeptorenblockers mit einem Kalziumantagonisten, gemessen am Rückgang der belastungsinduzierten ST-Streckensenkung, eine Verstärkung des antiischämischen Effekts auf. Auch die Kombination eines Nitrats, 120 mg ISDN retard mit dem Kalziumantagonisten Diltiazem in einer Dosis in 120 mg, erwies sich über einen Zeitraum von 12 h nicht nur gegenüber Placebo, sondern auch gegenüber der alleinigen Verabreichung beider Substanzen als stärker antiischämisch wirksam. Es ist anzunehmen, daß auch die Kombination eines Nitrats mit dem Kalziumantagonisten Verapamil eine verstärkte antiischämische Wirkung zur Folge hat (Lehmann et al. 1988, 1989).

Bezüglich der Reduktion der Häufigkeit von Angina pectoris und ischämischen Episoden im Langzeit-EKG ist durch eine 2fache Kombination ein Wirkungszuwachs zu erwarten, der gegenüber der jeweiligen Monotherapie in einer Größenordnung von 20–30 % liegt (Egstrup 1988; Strauss u. Parisi 1988; Kostuk u. Pflugfelder 1987; Findlay et al. 1986). Nur wenige Untersuchungen wurden zur Beantwortung der Frage durchgeführt, ob durch die Kombination aller 3 Substanzen mehr zu erreichen ist als mit einer 2fachen Kombination. Studien, die keinen zusätzlichen Therapiegewinn belegen, stehen Berichte gegenüber, die von einer Überlegenheit der 3fachen Kombination ausgehen. Schwierigkeiten in der Wertung ergeben sich, wenn zu befürchten ist, daß Nitrate in Dosierungen verabreicht wurden, die zur Toleranz führen, so daß möglicherweise nur der Wert einer 2fachen Kombination geprüft wurde (Tolins et al. 1984; Nesto et al. 1985; Stone et al. 1988).

In einer eigenen Studie wurde Patienten mit koronarer Herzerkrankung Metoprolol, ISDN sowie entweder der Kalziumantagonist Nisoldipin oder Diltiazem allein oder in Kombination verabreicht. Bei allen Formen der 2fachen Kombination ergab sich ein stärkerer Rückgang der belastungsinduzierten ST-Streckensenkung als unter alleiniger β-Rezeptorenblockergabe. Ein weiterer Wirkungszuwachs war bei einer 3fachen Kombination zu verzeichnen, wenn der Kalziumantagonist Diltiazem verabreicht wurde, nicht jedoch bei Verwendung des Kalziumantagonisten Nisoldipin. Die Effektivität der 3fachen Kombination resuliert dabei offenbar aus der synergistischen Wirkung von Nitrat und dem Kalziumantagonist Diltiazem, ein Synergismus, der zwischen Nitrat und dem Dihydropyridinderivat Nisoldipin nicht zu bestehen scheint (Lehmann et al. 1988).

Die vorliegenden Daten berechtigen somit zu der Annahme, daß bei der Mehrzahl der Patienten durch eine Mehrfachtherapie ein stärkerer antianginöser und antiischämischer Effekt erzielt wird. Sie erfordern es jedoch auch, den erwarteten Wirkungszuwachs individuell zu objektivieren.

Thrombozytenaggregationshemmer

Grundlage für den Einsatz von Thrombozytenaggregationshemmern bei koronarer Herzerkrankung stellt die Erkenntnis dar, daß der Verlauf der Koronarsklerose wesentlich durch das Auftreten von Plaquefissuren mit nachfolgenden intraintimalen und intraluminalen Thrombusbildungen bestimmt wird. Die umfangreichen Untersuchungen über den präventiven Wert von Thrombozytenaggregationshemmern beziehen sich wiederum überwiegend auf Patienten nach überstandenem Herzinfarkt. Einheitlich im Trend, in einigen Studien statistisch gesichert, ergab sich in diesen Sekundärpräventionsstudien eine Reduktion der Reinfarktrate, die gepoolten Daten zufolge mit 30 % anzusetzen ist (Antiplatelet Trialists' Collaboration 1988). In einer Primärpräventionsstudie, in der Ärzten Acetylsalicylsäure verabreicht wurde, zeigte sich bei unbeeinflußter Gesamtletalität eine deutlich verminderte Infarkthäufigkeit in dem mit Acetylsalicylsäure behandelten Kollektiv. Weitere Analysen ergaben, daß offenbar über 50jährige in besonderem Maße von Thrombozytenaggregationshemmern profitieren. Ob die Tatsache, daß die Sterblichkeit in dieser Studie insgesamt deutlich niedriger war als vorauskalkuliert wurde, die Aussagekraft der Ergebnisse begrenzt, bleibt offen. Zu beachten ist jedoch, daß hämorrhagische Hirninfarkte in der behandelten Gruppe signifikant häufiger auftraten. Eine zurückhaltende Handhabung des Einsatzes von Thrombozytenaggregationshemmern beim Vorliegen einer diabetischen Retinopathie oder unkontrollierter Hypertonie scheint damit berechtigt (Steering Committee 1989). Diese insgesamt günstigen Ergebnisse konnten jedoch in einer englischen Studie nicht bestätigt werden, wenngleich die Autoren einen positiven Effekt in einer Größenordnung zwischen 10 und 20% nicht ausschließen (Peto et al. 1988).

Die Vielzahl der vorliegenden Daten läßt es als vernünftig erscheinen, bei allen Patienten mit manifester koronarer Herzerkrankung, unabhängig vom Vorliegen eines abgelaufenen Infarkts, Acetylsalicylsäure in einer Dosis von 100–350 mg als Thrombozytenaggregationshemmer einzusetzen, sofern keine spezifischen Kontraindikationen bestehen.

Schlußfolgerungen

Nitrate, β-Rezeptorenblocker und Kalziumantagonisten, allein oder in Kombination verabreicht, sind hoch effektiv in der Behandlung der Myokardischämie. Ein individuell unterschiedliches Wirkungsausmaß, wobei die Variationen des Ansprechens auf Kalziumantagonisten am größten sind, erfordert es, nicht nur den Effekt einer Kombinationsbehandlung, sondern auch den einer Monotherapie zu objektivieren (Abb. 6).

Zum gegenwärtigen Zeitpunkt ist der Einfluß dieser Substanzen auf die Progression der Koronarsklerose nicht ausreichend belegt, wenngleich dies für Kalziumantagonisten als möglich angenommen werden kann.

Bezüglich der Beeinflussung der Morbidität und Letalität wurde für β-Rezeptorenblocker ein sekundärer präventiver Effekt gesichert, eine primäre präventive Wirkung ist zu vermuten. Ob und in welchem Umfang dies der antiischämischen Wirkung zuzuschreiben ist, bleibt offen. Subgruppenanalysen in Sekundärpräventionsstudien, die sich auf Patienten mit nichttransmuralem Infarkt beziehen, schließen den protektiven Effekt einer auf eine Ischämieverhinderung gerichteten Therapie zumindest nicht aus.

Eine antiischämische Medikation bedeutet damit zum gegenwärtigen Zeitpunkt eine deutliche Verbesserung der subjektiven Symptomatik, während der Ansatz, durch eine konsequente antiischämische Medikation die Prognose zu verbessern, nach wie vor hypothetisch bleibt. Er läßt sich jedoch damit begründen, daß u. U. prognostisch ungünstige Strukturveränderungen, die bei rezidivierender Myokardischämie auftreten, vermieden und u. U. lebensbedrohliche, ischämiebedingte Rhythmusstörungen verhindert werden. Basierend auf

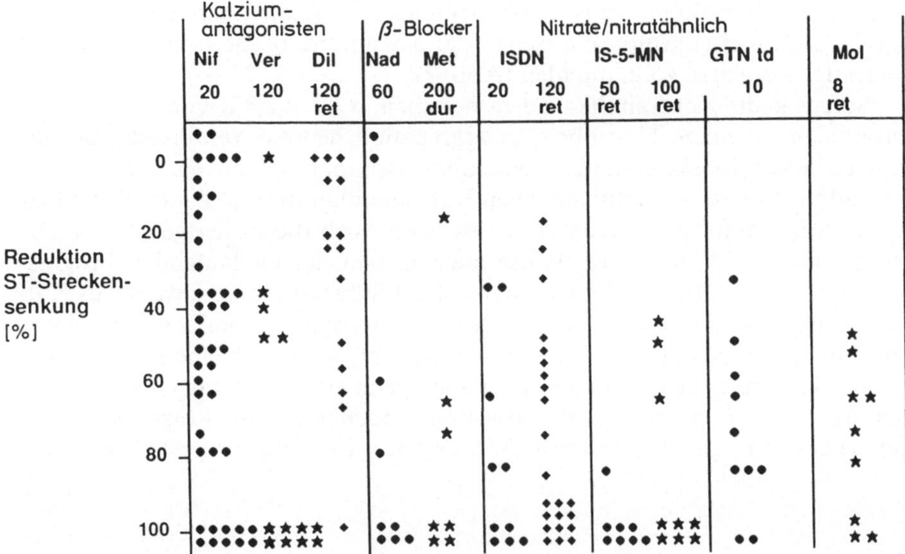

Abb. 6. Individuelles Ansprechen auf verschiedene antiischämisch wirksame Substanzen, beurteilt anhand der Reduktion der belastungsinduzierten ST-Streckensenkung (Angaben der Medikamentendosierung in mg; *Nif* Nifedipin, *Ver* Verapamil, *Dil* Diltiazem, *Nad* Nadolol, *Met* Metoprolol, *ISDN* Isosorbiddinitrat, *IS-5-MN* Isosorbid-5-mononitrat, *GTNtd* Glyceroltrinitrat transdermal, *Mol* Molsidomin)

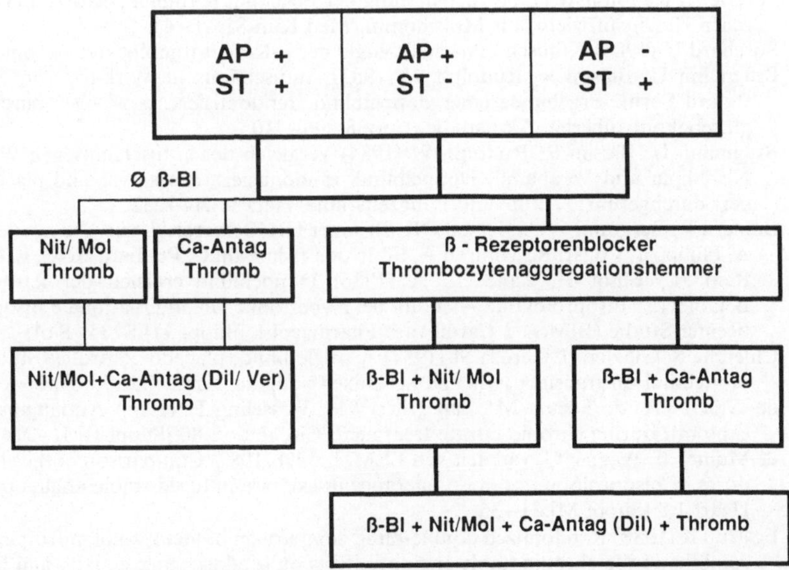

Abb. 7. Stufenplan zur Pharmakotherapie der Myokardischämie (*AP* Angina pectoris, *ST* ST-Streckensenkung, *Thromb* Thrombozytenaggregationshemmer, *Nit* Nitrate, *Mol* Molsidomin, *Ca-Antag* Kalziumantagonisten, *Ver* Verapamil, *Dil* Diltiazem, *β-Bl* β-Rezeptorenblocker)

den dargestellten Befunden ist bei Patienten mit Myokardischämie, unabhängig davon, ob eine symptomatische oder asymptomatische Ischämie besteht folgendes Vorgehen zu empfehlen (Abb. 7):

Sofern keine Kontraindikationen bestehen, wird ein β-Rezeptorenblocker zusammen mit einem Thrombozytenaggregationshemmer verabreicht. Bei unzureichendem Effekt erweitert man diese Behandlung um ein Nitrat bzw. Molsidomin oder alternativ um einen Kalziumantagonisten. Dieser Substanzklasse wäre dann der Vorzug zu geben, wenn sich die bisherigen Daten zur Beeinflussung der Koronarsklerose auch in den derzeit laufenden Progressionsstudien bestätigen. Medikamente aller 3 Substanzklassen, wobei als Kalziumantagonist Diltiazem vorzuziehen ist, werden zusammen mit einem Thrombozytenaggregationshemmer verabreicht, wenn sich eine 2fache Kombinationstherapie nicht als ausreichend wirksam erweist. Parallel mit der Einleitung der medikamentösen Therapie muß jedoch immer die Frage nach anderen Therapiemöglichkeiten wie PTCA und Bypassoperation gestellt werden.

Literatur

Ablad B, Björkman J-A, Gustafsson D, Hansson G, Östlund-Lindqvist A-M, Pettersson K (1988) The role of sympathic activity in atherogenesis: Effects of β-blockade. Am Heart J 116:322–327

Antiplatelet Trialists Collaboration (1988) Secondary prevention of vascular disease by prolonged antiplatelet treatment. Br Med J 296:320–331

Beyerle A, Rudolph W (1986) Behandlung der Belastungs-Angina pectoris und der chronischen Herzinsuffizienz mit Molsidomin. Med Klin 81:41–44

Borchard U (1988) Klinische Pharmakologie der β-Rezeptorenblocker. Aesopus, Basel

Brügmann U, Blasini R, Rudolph W (1983) Antiischämische Wirkung von Nifedipin in Retard-Form. Ergebnisse einer doppelblind, randomisiert, cross-over durchgeführten, placebokontrollierten Akutstudie. Herz 8:206–210

Brügmann U, Blasini R, Rudolph W (1984) Vergleich der antiischämischen Wirkung von Nisoldipin und Verapamil. Doppelblind, randomisiert, cross-over und placebokontrolliert durchgeführte Akut- und Langzeitstudie. Herz 9:244–252

Bühler FR, Berglund G, Anderson OK, Brunner HR, Scherrer U, van Brummelen P, Distler A, Philipp T, Fogari R, Mimran A, Fourcade J, dalPalu C, Prichard BNC, Backhouse CI, Reid JL, Elliot H, Zanchetti A (1986) Doppelblindvergleich der kardioselektiven Betablocker Bisoprolol und Atenolol bei Hypertonie: Die internationale Bisoprolol-Multicenter-Studie (BIMS). J Cardiovasc Pharmacol 8 [Suppl 11]:S135–S141

Chiercha S, Glazier JJ, Gerosa St (1987) A single-blind, placebo-controlled study of effects of atenolol on transient ischemia in "mixed" angina. Am J Cardiol 60:36A–40A

de Graeff PA, de Leeuw MJ, van Gilst WH, Wesseling H (1989) Antianginal effects of captopril during chronic nitrate treatment. Circulation 80 [Suppl II]:II–214

de Muinck E, Wagner G, van den Ven LLM, Lie KI (1987) Comparison of the effects of two doses of bisoprolol on exercise tolerance in exercise-induced stable angina pectoris. Eur Heart J 8 [Suppl M]:31–35

Egstrup K (1988) Randomized double-blind comparison of metoprolol, nifedipine, and their combination in chronic stable angina: Effects on total ischemic activity and heart rate at onset of ischemia. Am Heart J 116:971–978

Findlay IN, MacLeod K, Ford M, Gillen G, Elliott A, Dargie HJ (1986) Treatment of angina pectoris with nifedipine and atenolol: Efficacy and effect on cardiac function. Br Heart J 55:240–245

Fitzgerald JD (1987) By what means might beta blockers prolong life after acute myocardial infarction? Eur Heart J 8:945–951
Freedman SB, Jamal SM, Harris PJ, Kelly DT (1987) Comparison of carvedilol and atenolol for angina pectoris. Am J Cardiol 60:499–502
Frithz G, Weiner L (1986) Langzeitwirkungen von Bisoprolol auf Blutdruck, Serumlipide und HDL-Cholesterin bei Patienten mit essentieller Hypertonie. J Cardiovasc Pharmacol 8 [Suppl 11]:S149–S153
Gibson RS (1989) Current status of calcium channel-blocking drugs after Q wave and non-Q wave myocardial infarction. Circulation 80 [Suppl IV]:IV-107–IV-119
Herrmann JM, Bischof F, von Heymann F, Freischuetz G, Burghagen H (1988) Effects of celiprolol on serum lipids in systemic hypertension. Am J Cardiol 61:41C–44C
Hogan JC, Lewis MJ, Henderson AH (1989) Failure of N-acetylcysteine (NAC) to prevent nitrate tolerance in patients with angina. Eur Heart J 10[Abstr Suppl]:153
Hung J, Lamb IH, Connolly StJ, Jutzy KR, Goris ML, Schroeder JS (1983) The effect of diltiazem and propranolol, alone and in combination, on exercise performance and left ventricular function in patients with stable effort angina: a double-blind, randomized, and placebo-controlled study. Circulation 68:560–567
Imperi GA, Lambert CR, Coy K, Lopez L, Pepine CJ (1987) Effects of titrated beta blockade (metoprolol) on silent myocardial ischemia in ambulatory patients with coronary artery disease. Am J Cardiol 60:519–524
Jackson G, Schwartz J, Kates RE, Winchester M, Harrison DC (1980) Atenolol: Once-daily cardioselective beta blockade for angina pectoris. Circulation 61:555–560
Kostuk WJ, Pflugfelder P (1987) Comparative effects of calcium entry-blocking drugs, β-blocking drugs, and their combination in patients with chronic stable angina. Circulation 75 [Suppl V] V-114–V-121
Lahiri A, Rodrigues EA, DasGupta P, Jain D, van der Does R, Raftery EB (1989) Effects on carvedilol in patients with impaired left ventricular function due to ischaemic heart disease. Z Kardiol 78 [Suppl 3]:21–27
Lehmann G, Reiniger G, Dirschinger J, Rudolph W (1988) Effectiveness on antianginal substances: Mono-, double- or triple-therapy? Eur Heart J 9 [Abstr Suppl 1] 157
Lehmann G, Reiniger G, Dirschinger J, Rudolph W (1989) Enhanced effectivity by combination of nitrates and the calcium antagonist diltiazem in patients with angina pectoris. Eur Heart J 10 [Abstr Suppl] 154
Lichtlen PR, Hugenholtz P, Rafflenbeul W, Jost S, Hecker H, INTACT-Study Group (1989) Retardation of the progression of coronary artery disease with nifedipine. Results of INTACT. Circulation 80 [Suppl II]:II-382
Loaldi A, Polese A, Montorsi P, deCesare N, Fabbiocchi F, Ravagnani P, Guazzi MD (1989) Comparison of nifedipine, propranolol and isorbide dinitrate on angiographic progression and regression of coronary arterial narrowings in angina pectoris. Am J Cardiol 64:433–439
Lynch P, Dargie H, Krikler S, Krikler D (1980) Objective assessment of antianginal treatment: a double-blind comparison of propranolol, nifedipine, and their combination. Br Med J 281:184–187
McDevitt DG (1988) Medikamentös bedingte Störung der Sexualfunktion. Ein vermeidbares Problem. Nr. 3, Beta-Blocker und sexuelle Dysfunktion. ADIS, Manchester
Miller NE (1987) Effects of adrenoceptor-blocking drugs on plasma lipoprotein concentrations. Am J Cardiol 60:17E–23E
Nesto RW, White HD, Ganz P, Koslowski J, Wynne J, Holman BL, Antman E (1985) Addition of nifedipine to maximal beta-blocker-nitrate therapy: Effects on exercise capacity and global left ventricular performance at rest and during exercise. Am J Cardiol 55:3E–8E
Olsson G, Rehnqvist N (1987) Postinfarction metoprolol treatment: Effects on prognosis in relation to serum cholesterol concentrations. Cardiology 74:457–464
Packer M, Lee WH, Kessler PD, Gottlieb StS, Medina N, Yushak M (1987) Prevention and reversal of nitrate tolerance in patients with congestive heart failure. N Engl J Med 317:799–804

Packer M (1989a) Second generation calcium channel blockers in the treatment of chronic heart failure: Are they any better than their predecessors? JACC 14:1339–1342

Packer M (1989b) Combined beta-adrenergic and calcium-entry blockade in angina pectoris. N Engl J Med 320:709–718

Parker JO, Farrell B, Lahey KA, Rose BF (1987) Nitrate tolerance: the lack of effect of N-acetylcysteine. Circulation 76:572–576

Peto R, Gray R, Collins R, Wheatley K, Hennekens C, Jamrozik K, Warlow C, Hafner B, Thompson E, Norton S, Gilliland J, Doll R (1988) Randomised trial of prophylactic daily asperin in British male doctors. Br Med J 296:313–316

Prichard BNC, Tomlinson B (1989) Carvedilol – A double pronged attack on angina. Clinical pharmacology of multiple action compounds. Z Kardiol 78[Suppl 3]:1–6

Reiniger G, Rudolph W (1985) Therapie der koronaren Herzerkrankung mit Nitroglycerin-Pflastern. Verhalten der antiischämischen Wirksamkeit bei kontinuierlicher Verabreichung und nach „pflasterfreiem" Intervall. Herz 10:305–311

Reiniger G, Kraus F, Dirschinger J, Blasini R, Rudolph W (1985) Hochdosierte transdermale Nitroglycerin-Therapie: Wirkungsverlust innerhalb von 24 Stunden? Herz 10:157–162

Reiniger G, Menke G, Boertz A, Kraus F, Rudolph W (1987) Intervalltherapie zur effektiven Behandlung der Angina pectoris mit Nitroglycerin-Pflastersystemen. Kontrollierte Studie mit Bestimmung der Nitroglycerin-Plasmaspiegel. Herz 12:68–73

Roberts WC (1987) Blood lipid levels and antihypertensive therapy. Am J Cardiol 60:30E–35E

Rudolph W, Blasini R (1984) Nitrattoleranz: ein klinisch bedeutsames Problem? Herz 9:115–122

Rudolph W, Dirschinger J (1984) Pharmakotherapie der Angina pectoris. In: Hombach V (Hrsg) Kardiologie. Bd 1: Die koronare Herzkrankheit, Schattauer, Stuttgart New York, S 173–204

Rudolph W, Dirschinger J (1985) Effectiveness of molsidomine in the longterm treatment of exertional angina pectoris and chronic congestive heart failure. Am Heart J 109:670–674

Rudolph W, Blasini R, Froer KL, Brügmann U, Mannes A, Hall D (1981) Effects of acute and chronic administration of isosorbide dinitrate, sustained-release form, in patients with angina pectoris. In: Lichtlen PR, Engel H-J, Schrey A, Swan HJC (eds) Nitrates III-Cardiovascular effects. Springer, Berlin Heidelberg New York, pp 76–81

Rudolph W, Blasini R, Reiniger G, Brügmann U (1983) Tolerance development during isosorbide dinitrate treatment: Can it be circumvented? Z Kardiol 72 [Suppl 3]:195–198

Rudolph W, Dirschinger J, Blasini R, Reiniger G, Kraus F (1985) Nitrattoleranz: Wann tritt sie auf? Wie ist sie vermeidbar? Dtsch Ärzteblatt 82:3421–3432

Rudolph W, Dirschinger J, Reiniger G, Beyerle A, Hall D (1988) When does nitrate tolerance develop? What dosages and which intervals are necessary to ensure maintained effectiveness? Eur Heart J 9[Suppl A]:63–72

Rudolph W, Dirschinger J, Hall D, Reiniger G, Beyerle A (1989) Effectiveness of interval nitrate therapy in angina pectoris. Eur Heart J 10[Suppl A]:50–55

Smith PJ, Talbert RL (1986) Sexual dysfunction with antihypertensive and antipsychotic agents. Clin Pharm 5:373–384

Steering Committee of the Physicians Health Study Research Group (1989) Final report on the asperin component of the ongoing physicians health study. N Engl J Med 321:129–135

Stone PH, Ware JH, deWood MA, Gore JM, Eich RH, Pietro DA, Parisi AF, Nesto RW, Boden WE, Sharma SC, Vlay StC, Ennis LE, Gianelly RE, Turi ZG, McCall NT, Curtis DG, Chierchia S, Maseri A, Braunwald E (1988) The efficacy of the addition of nifedipine in patients with mixed angina compared to patients with classic exertional angina: A multicenter, randomized, double-blind, placebo-controlled clinical trial. Am Heart J 116:961–970

Strauss WE, Parisi AF (1988) Combined use of calcium-channel and beta-adrenergic blockers for the treatment of chronic stable angina. Rationale, efficacy, and adverse effects. Ann Intern Med 109:570–581

Tolins M, Weir E, Chesler E, Pierpont GL (1984) "Maximal" drug therapy is not necessarily optimal in chronic angina pectoris. JACC 3:1051–1057

Uusitalo AJ, Keyriläinen O (1979) Slow-release metoprolol in angina pectoris. A comparative study of a cardioselective β-blocking drug, metoprolol, in ordinary and slow-release tablets (durules) in the treatment of angina pectoris. Ann Clin Res 11:199–204

Weiner DA, Ryan TJ, McCabe CH, Luk St, Chaitman BR, Sheffield LT, Tristani F, Fisher LD (1987) Significance of silent myocardial ischemia during exercise testing in patients with coronary artery disease. Am J Cardiol 59:725–729

Chronopharmakologie antiischämisch wirksamer Arzneimittel

B. Lemmer

Alle physiologischen Funktionen des Menschen unterliegen einer ausgeprägten zirkadianen Rhythmik (Übersicht siehe Lemmer 1984; Lemmer 1989). Dies gilt auch für das kardiovaskuläre System, in dem vielfach Rhythmen nachgewiesen wurden; z.B. in systolischem und diastolischem Blutdruck, Herzfrequenz, Parametern des EKG, peripherem Widerstand, Organdurchblutung, Blutviskosität und -aggregabilität, vaskulärer Reagibilität und Barorezeptorenreflex, in den Plasmakonzentrationen des Neurotransmitters Noradrenalin und des „second messenger" cAMP (Abb. 1), bis hinunter auf die zelluläre und subzelluläre Ebene des sympathischen Nervensystems, d.h. das β-Adrenozeptor-Effektor-System und die cAMP-vermittelte Signalübertragung (Lemmer 1987, 1988, 1989, Lemmer et al. 1987). Gut dokumentiert ist auch, daß pathologische kardiovaskuläre Ereignisse wie Herzinfarkt, Hirn-

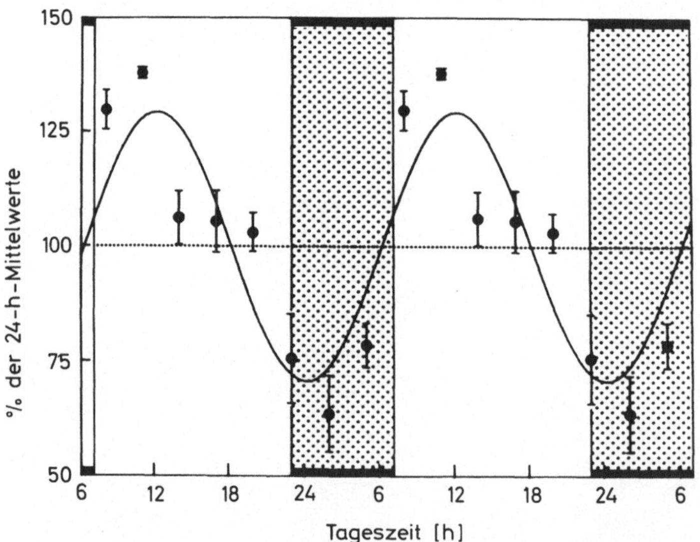

Abb. 1. Zirkadianer Rhythmus im cAMP-Gehalt des menschlichen Plasmas. Dargestellt sind Mittelwerte ± SEM von 6 gesunden Probanden als Prozent der individuellen 24-h-Mittelwerte, die *durchgezogene Linie* gibt die Anpassung an eine Sinusfunktion wieder, doppelte Darstellung zur Verdeutlichung des Rhythmus. Die individuellen 24-h-Mittelwerte des cAMP-Gehalts liegen im Bereich von 8–14 pmol/ml Plasma. (Aus Lemmer 1988)

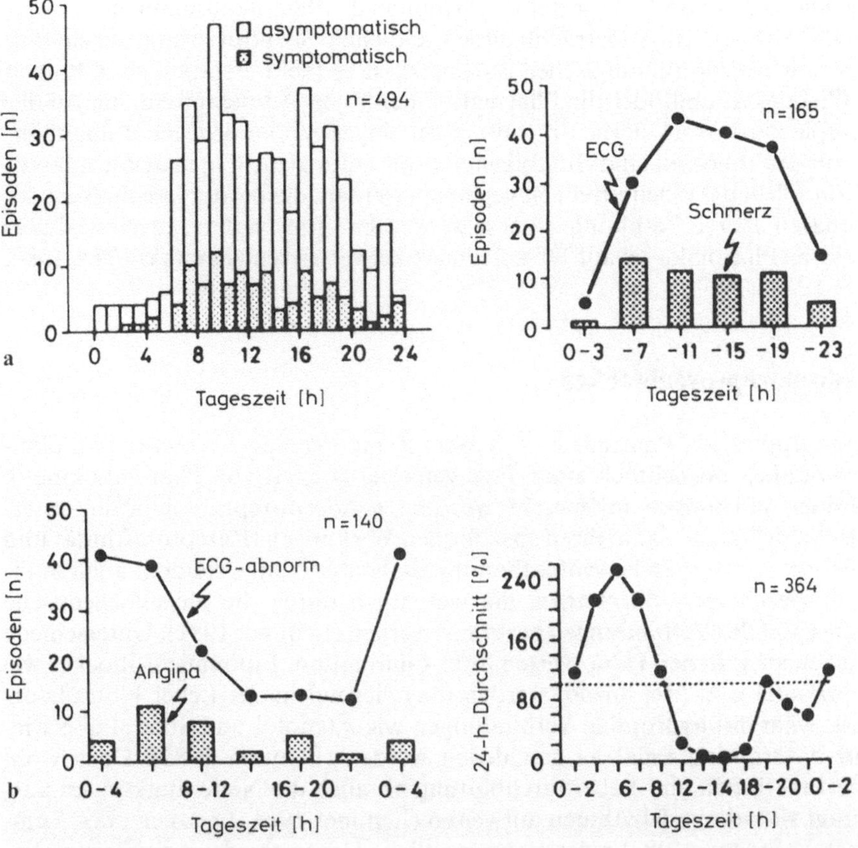

Abb. 2. Zirkadiane Rhythmen in pathophysiologischen Ereignissen bei Angina pectoris. **a** Zirkadiane Verteilung von 382 asymptomatischen (ST-Streckensenkung) und 112 symptomatischen Ischämieepisoden bei Patienten mit stabiler Angina pectoris (*links*; nach Hausmann et al. 1987); ST-Streckensenkung und schmerzhafte Episoden (n = 165) im ambulatorischen 24-h-EKG bei Patienten, bei denen anschließend eine Koronarangiographie durchgeführt wurde (*rechts*; nach v. Arnim et al. 1985). **b** ST-Streckenerhöhung bei 25 Patienten mit koronarspastischer Angina (*rechts*; nach Araki et al. 1983); Angina-pectoris-Anfälle und EKG-Veränderungen wie ST-Streckenerhöhung, ST-Streckendepression oder T-Wellen-Pseudonormalisierung bei 13 Patienten mit koronarspastischer Angina (*links*; Waters et al. 1984)

infarkt, Angina-pectoris-Anfälle bzw. stumme Ischämien nicht gleichmäßig häufig zu allen Tageszeiten innerhalb von 24 h auftreten (Araki et al. 1983; v. Arnim et al. 1985; Joy et al. 1982; Marshall 1977; Mitler et al. 1987; Muller et al. 1985; siehe Lemmer 1987, 1988, 1989). Während bei vasospastischer Angina pectoris Anfallshäufigkeit und ST-Streckenerhöhungen ihre Maxima in den nächtlichen Stunden um 4–6 Uhr haben (Araki et al. 1983; v. Arnim et al. 1985; Waters et al. 1984; Yasue et al. 1979), treten bei stabiler Angina pectoris symptomatische und asymptomatische Ischämieepisoden am häufigsten zwischen 8 und 12 Uhr auf (v. Arnim et al. 1985; Hausmann et al. 1987; s. auch Abb. 2). In Anbetracht dieser zeitlichen Strukturierung physiologischer und pathophysiologischer Ereignisse ist es nicht erstaunlich, daß auch die Wirkungen und/oder die Pharmakokinetik von Arzneimitteln, die bei der Therapie kardiovaskulärer Erkrankungen eingesetzt werden, nicht unabhängig von der Tageszeit sind. Im folgenden soll nur auf die Chronopharmakologie von solchen Arzneimitteln eingegangen werden, die bei der Behandlung der koronaren Herzerkrankung eingesetzt werden. Für andere kardiovaskulär wirksame Pharmaka sei auf Übersichtsartikel verwiesen (Lemmer 1984, 1987, 1988, 1989).

β-Adrenozeptorenblocker

Diese Gruppe von Pharmaka ist in einer Reihe tierexperimenteller und klinischer Studien hinsichtlich einer Tageszeitabhängigkeit von Pharmakokinetik und/oder Wirkungen untersucht worden. *β*-Adrenozeptorenblocker unterscheiden sich nicht nur in ihren spezifischen Wirkungen (Rezeptoraffinität und -selektivität, intrinsische sympathomimetische Aktivität), sondern auch in ihren unspezifischen Wirkungen, die wesentlich durch die physikochemische Eigenschaft der Verbindungen bestimmt werden (Lemmer 1982). Unterschiede bestehen auch in den Hauptwegen ihrer Elimination. Lipophile *β*-Blocker wie Propranolol und Metoprolol werden überwiegend in der Leber biotransformiert, während hydrophile Verbindungen wie Atenolol und Sotalol überwiegend unverändert renal ausgeschieden werden (Lemmer 1982). Da sowohl Leberfunktionen und Leberdurchblutung als auch die Nierenfunktionen ausgeprägt zirkadiane Rhythmen aufweisen (Lemmer 1984; Lemmer 1989; Lemmer u. Labrecque 1987) müssen tageszeitliche Unterschiede in der Pharmakokinetik der *β*-Blocker in Betracht gezogen werden. Ausführliche Daten zur Chronopharmakokinetik verschiedener *β*-Blocker liegen nur aus tierexperimentellen Untersuchungen vor, beim Menschen wurde bisher nur Propranolol untersucht.

Bei nachtaktiven Ratten konnten für Propranolol, Metoprolol, Atenolol und Sotalol – und damit unabhängig von einer überwiegend hepatischen oder renalen Elimination – signifikant kürzere Eliminationshalbwertszeiten in der Aktivitätsperiode in der Nacht als in der Ruheperiode am Tag nachgewiesen werden (Lemmer u. Bathe 1982; Lemmer et al. 1983, 1985). Ebenso war die herzfrequenzsenkende Wirkung von Propranolol bei wachen Ratten in der Phase des erhöhten Symphatikustonus in der Nacht stärker als in der Ruheperiode am Tag (Lemmer et al. 1983, 1985).

Im Prinzip gleiche Ergebnisse wurden bei gesunden Probanden erhalten, berücksichtigt man, daß im Gegensatz zur Ratte der Mensch ein tagaktives Lebewesen mit einem erhöhten Sympathikustonus in seiner Aktivitätsperiode am Tag ist. Letzteres spiegelt sich im zirkadianen Rhythmus im cAMP-Gehalt im Plasma wider (Abb. 1). Bei diesen Probanden waren nach orale Gabe von Propranolol zu 4 verschiedenen Tageszeiten signifikante tageszeitabhängige

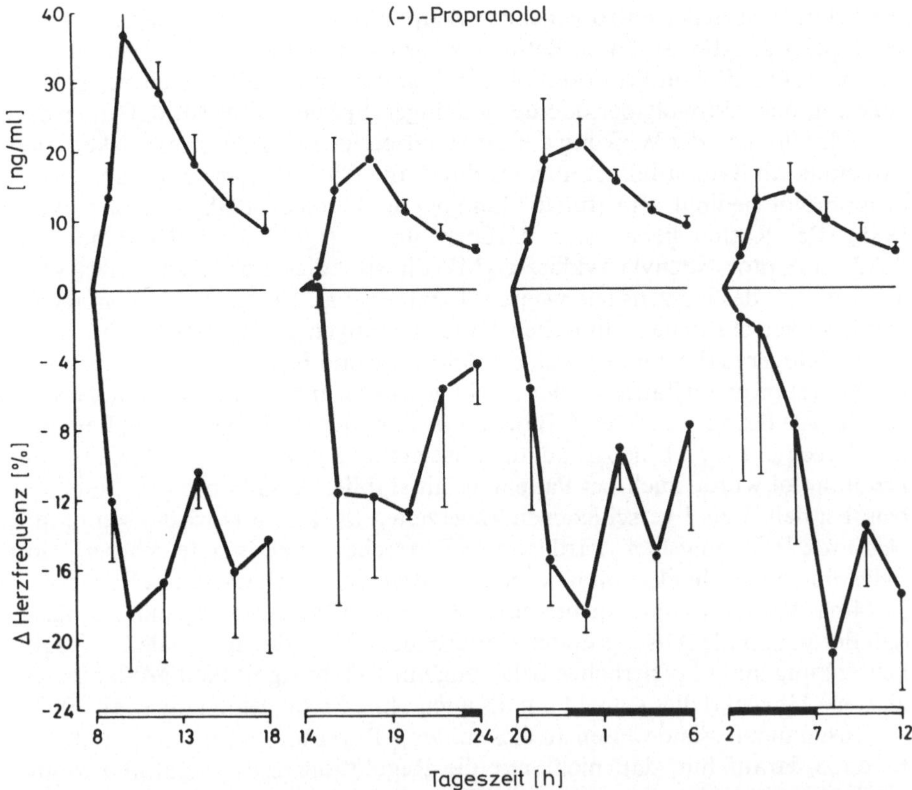

Abb. 3. Chronopharmakologie von Propranolol nach oraler Gabe von 80 mg des Razemates zu 4 verschiedenen Tageszeiten (8, 14, 20, 2 Uhr) mit jeweils einer Woche Auswaschphase. Plasmakonzentrationen an (–)-Propranolol (*oben*) und prozentuale Senkung der Herzfrequenz im Sitzen (*unten*) im Vergleich zu den zirkadianen Kontrollwerten. Die Ruheperiode ist durch den *schwarzen Balken* gekennzeichnet. (Nach Langner u. Lemmer 1988)

Unterschiede in pharmakokinetischen Parametern nachzuweisen (Langner u. Lemmer 1988): Bei morgendlicher Einnahme von Propranolol um 8 Uhr waren maximale Plasmakonzentration (C_{max}) und AUC am größten, Resorptionszeit (t_{max}) und Eliminationshalbwertszeit am kürzesten (Abb. 3). Interessanterweise wies der stereospezifische Metabolismus von Propranolol beim Menschen keine Tageszeitabhängigkeit auf (Langner u. Lemmer 1988). Intensität und Verlauf des herzfrequenzsenkenden Effekts von Propranolol waren tageszeitabhängig, korrelierten jedoch nicht mit den chronopharmakokinetischen Befunden. Bei Einnahme von Propranolol um 2 Uhr nachts wurde die Herzfrequenz in den ersten 4 h nicht wesentlich gesenkt; 2 h später hingegen war mit Anstieg des Sympathikustonus der gleiche frequenzsenkende Effekt zu beobachten wie nach Propranololeinnahme um 8 Uhr morgens (Abb. 3), wobei in jeder Studie der Effekt jeweils im Sitzen im Vergleich zu den zirkadianen Kontrollwerten berechnet wurde. Die (–)-Propranolol-konzentrationen hingegen differierten zu den entsprechenden Zeitpunkten maximaler Wirkungen um den Faktor 3 (Abb. 3). Diese Befunde zeigen, daß bei Ratte und Mensch die Pharmakokinetik von Propranolol eine Tagesrhythmik mit schnellerer Elimination in der Aktivitätsperiode der jeweiligen Spezies aufweist, daß aber die Tagesrhythmik in der Wirkung auf die Herzfrequenz eher durch die zirkadiane Rhythmik im Sympathikustonus als durch die Chronopharmakokinetik von Propranolol bedingt sein dürfte (Langner u. Lemmer 1988; Lemmer et al. 1985). Bei Ratten ließen sich darüber hinaus signifikante Rhythmen im β-Adrenozeptor-Adenylatzyklase-cAMP-Phosphodiesterase-System des Vorderhirns und des Herzens nachweisen (Lemmer et al. 1987). Übereinstimmend wurde in verschiedenen klinischen Untersuchungen gezeigt, daß selbst nach mehrwöchiger Gabe von β-Blockern Blutdruck und Herzfrequenz von Hypertonikern stärker am Tag als in der Nacht gesenkt wurden, und dies unabhängig von der Art des verwendeten β-Blockers (s. Lemmer 1987, 1988, 1989; Lemmer u. Labrecque 1987). Eine zirkadiane Phasenabhängigkeit in der Wirkung von Propranolol wurde auch bei Patienten mit stabiler Angina pectoris beschrieben, bei denen zu 3 verschiedenen Tageszeiten (8, 12, 16 Uhr) die symptombegrenzte Belastungstachykardie und ST-Streckensenkung unter körperlicher Belastung ohne Medikation bzw. nach mehrwöchiger Therapie mit Propranolol (4mal 40 mg/kg) untersucht wurde (Joy et al. 1982). Die Ergebnisse zeigen, daß der maximale Anstieg in der Herzfrequenz bzw. die maximale ST-Streckensenkung unter körperlicher Belastung um 16 Uhr signifikant größer waren als um 8 Uhr, und dies sowohl mit als auch ohne Medikation (Joy et al. 1982).

Zusammenfassend weisen all diese unter β-Rezeptorenblockade erhaltenen Befunde darauf hin, daß nicht nur die Regulation des Sympathikustonus, sondern auch die kardiale und vaskuläre Ansprechbarkeit auf β-Rezeptorenblockade einer zirkadianen Rhythmik unterliegen müssen. Darauf ließen auch die bereits oben erwähnten tierexperimentellen Befunde schließen.

Kalziumkanalblocker

Ähnlich wie die β-Blocker beeinflussen auch Kalziumkanalblocker (Kalziumantagonisten) bei chronischer Gabe Blutdruck und Herzfrequenz von Hypertonikern unterschiedlich stark zu verschiedenen Tageszeiten (Gould et al. 1982a, b; White et al. 1985; siehe Lemmer 1987, 1988, 1989). Kalziumkanalblocker vom Verapamil-Typ und Diltiazem haben stärkere kardiale, negativ chronotrope Wirkungen, die vom Dihydropyrimidintyp wie Nifedipin und Nitrendipin einen dominanteren vasodilatierenden Effekt, so daß die Herzfrequenz reflektorisch ansteigen kann. Chronopharmakologische Untersuchungen bei Hypertonikern haben nun gezeigt, daß – ähnlich wie bei β-Blockern – die blutdrucksenkenden Wirkungen der Kalziumkanalblocker Verapamil und Nifedipin am Tag stärker als in der Nacht waren (Abb. 4). Die etwas unterschiedlichen Angriffspunkte der Kalziumkanalblocker spiegeln sich auch in ihren Wirkungen auf den zirkadianen Rhythmus in der Herzfrequenz wider: Während Verapamil die Herzfrequenz sowohl am Tag als auch in der Nacht –

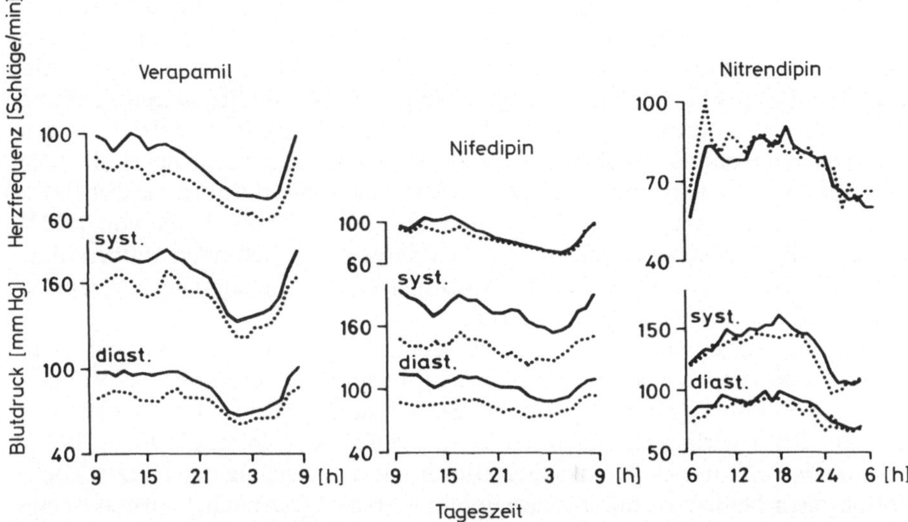

Abb. 4. Zirkadiane Phasenabhängigkeit der Wirkungen der Kalziumkanalblocker Verapamil (Gould et al. 1982a, b), Nifedipin (Gould et al. 1982a) und Nitredipin (White et al. 1985) auf zirkadiane Rhythmen in Blutdruck und Herzfrequenz von Hypertonikern vor (—) und nach (· · · ·) chronischer Arzneimittelgabe

wenn auch geringer – senkte, beeinflußte Nifedipin die Frequenz im 24-h-Tagesgang nicht, Nitrendipin erhöhte sie sogar in den frühen Morgenstunden (Abb. 4). Schließlich ist auch die Pharmakokinetik von Verapamil (Hla et al. 1988) und Nifedipin (Lemmer et al. 1989a) nicht unabhängig von der Tageszeit, wobei morgens C_{max} höher und/oder t_{max} kürzer sind als zu anderen Einnahmezeitpunkten.

Organische Nitrate

Es wurde bereits darauf hingewiesen, daß Angina-pectoris-Anfälle und stumme Ischämien nicht gleich häufig zu verschiedenen Tageszeiten auftreten, wobei ST-Streckenerhöhungen und ST-Streckensenkungen zu etwas unterschiedlichen Tageszeiten ihre Maxima haben (s. Abb. 2). Wiederum überrascht es nicht, daß die antianginöse Wirkung organischer Nitrate ebenfalls eine zirkadiane Phasenabhängigkeit aufweist. Bereits 1979 haben Yasue et al. bei Patienten mit einer Prinzmetal-Angina gezeigt, daß eine körperliche Belastung am Morgen einen Anfall mit einer entsprechenden ST-Streckenerhöhung auslöste, während eine sogar relativ stärkere Belastung am Nachmittag weder Anfall noch EKG-Veränderungen bewirkte. Glyceroltrinitrat morgens gegeben verhinderte nicht nur den belastungsinduzierten Angina-pectoris-Anfall, sondern führte zu einer Erweiterung der großen Koronararterien, die gleiche Dosis nachmittags gegeben hatte nur einen geringen Effekt (Yasue et al. 1979).

Zur Chronopharmakologie von Isosorbid-dinitrat (ISDN) und Isosorbid-5-mononitrat (IS-5-MN) liegen bisher nur Ergebnisse an gesunden Probanden vor (Lemmer et al. 1989 b; Scheidel et al. 1987, 1988). In einer Studie mit ISDN-Gabe zu 4 verschiedenen Tageszeiten (2, 8, 14, 20 Uhr) waren Senkung des systolischen Blutdrucks und reflektorische Erhöhung der Herzfrequenz am stärksten nach der nächtlichen Gabe von ISDN um 2 Uhr (Scheidel et al. 1988; unveröffentlichte Ergebnisse). In der Nacht ist nicht nur die normale Orthostasereaktion am ausgeprägtesten, sondern v.a. die auf orale Gabe von ISDN (Abb. 5). Zusammen mit den weiter oben dargelegten Befunden stützen diese Ergebnisse die Hypothese, daß das kardiovaskuläre System einer ausgeprägten zirkadianen Organisation unterliegt.

Tageszeitliche Variationen wurden auch in der Pharmakokinetik eines schnellfreisetzenden Präparats von IS-5-MN nachgewiesen, wobei bei Einnahme um 6.30 Uhr t_{max} 0,9 h betrug, bei Einnahme um 18.30 Uhr hingegen war t_{max} auf 2,1 h verlängert (Lemmer et al. 1989b; Scheidel et al. 1987, 1988). Maximale Senkung im systolischen Blutdruck und Anstieg der Herzfrequenz traten nach beiden Applikationszeiten etwa 0,6–1,0 h nach Pharmakoneinnahme auf, was auf unterschiedliche Dosis-Wirkungs-Beziehungen zu verschiedenen Tageszeiten hinweist (unveröffentlichte Ergebnisse). Hingegen war nach oraler Gabe eines Retardpräparates von IS-5-MN (Gabe um 8 Uhr oder 20 Uhr) keine Tageszeitabhängigkeit in der Pharmakokinetik festzustellen (Lemmer et al. 1988 b; unveröffentlichte Ergebnisse). Jedoch traten maximale

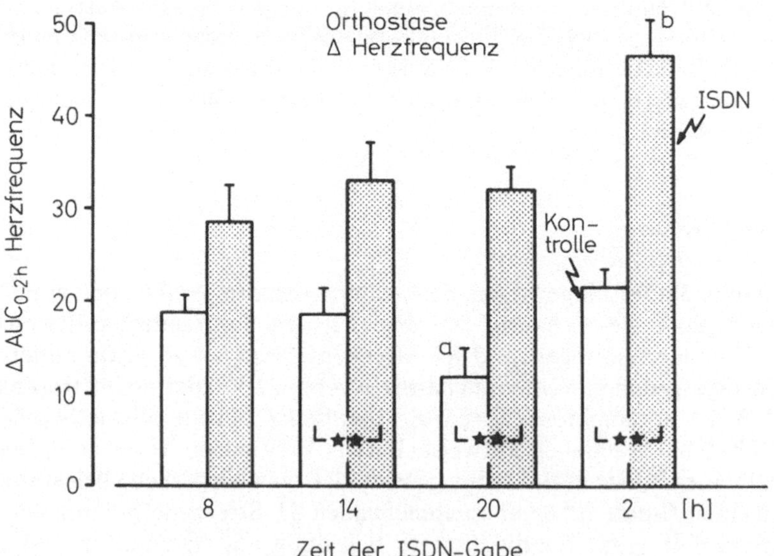

Abb. 5. Zirkadiane Veränderungen in der Orthostasereaktion (Zunahme der Herzfrequenz als AUC_{0-2h}. □ Keine Gabe von ISDN; ▨ bzw. nach oraler Gabe von 20 mg ISDN um 8, 14, 20 oder 2 Uhr bei gesunden Probanden. ANOVA-control und ISDN: $p<0{,}05$; Newman-Keuls a, b: $p<0{,}05$; ** control zu ISDN; $p<0{,}01$. (Unveröffentlicht)

Blutdrucksenkung und Frequenzsteigerung am Morgen zeitgleich mit den maximalen Plasmaspiegeln auf, während abends die maximalen Effekte etwa 2 h vor den maximalen Plasmaspiegeln von IS-5-MN gemessen wurden (Lemmer et al. 1989b; Scheidel et al. 1988). Über das bisher Gesagte hinaus zeigen diese klinischen Ergebnisse, daß auch die galenische Zubereitung ein unterschiedliches pharmakokinetisches Profil zu verschiedenen Tageszeiten nach sich ziehen kann.

Zusammenfassung

Klinische und tierexperimentelle Untersuchungen zeigen eindeutig, daß das kardiovaskuläre System und das Auftreten von Symptomen kardiovaskulärer Erkrankungen wie Herzinfarkt und Angina pectoris eine ausgeprägte zirkadiane Strukturierung aufweisen. Für antianginös wirksame Arzneimittel wie β-Adrenozeptorenblocker, Kalziumkanalblocker und organische Nitrate sind ebenfalls tageszeitabhängige Unterschiede in ihrer Pharmakokinetik und/oder ihren Wirkungen nachweisbar. Bei Arzneimittelprüfung, Stellung der Diagnose und Pharmakotherapie sind somit der Faktor „Tageszeit" bzw. zirkadiane Rhythmen zu berücksichtigen.

Literatur

Araki H, Koiwaya Y, Nakagaki O, Nakamura M (1983) Diurnal distribution of ST-segment elevation and related arrhythmias in patients with variant angina: a study by ambulatory ECG monitoring. Circulation 67:995–1000

Arnim T von, Höfling B, Schreiber M (1985) Characteristics of episodes of ST elevation or ST depression during ambulatory monitoring in patients subsequently undergoing coronary angiography. Br Heart J 54:484–488

Gould BA, Hornung RS, Mann S, Balasubramanian V, Raftery EB (1982a) Slow channel inhibitors verapamil and nifedipine in the management of hypertension. J Cardiovasc Pharmacol 4:5369–5373

Gould BA, Mann S, Kieso H, Balasubramanian V, Raftery EB (1982b) The 24-hour ambulatory blood pressure profile with verapamil. Circulation 65:22–27

Hausmann D, Nikutta P, Hartwig C-A, Daniel WG, Wenzlaff P, Lichtlen PR (1987) ST-Strecken-Analyse im 24-h-Langzeit-EKG bei Patienten mit stabiler Angina pectoris und angiographisch nachgewiesener Koronarsklerose. Z Kardiol 76:554–562

Hla KK, Henry JA, Phull R, Volane QN, Latham AN (1988) Chronopharmacology of verapamil. (25th Anniversary Int Symp Calcium Antagonists in Hypertension, Basel, 11.–12.2. 1988)

Joy M, Pollard CM, Nunan TO (1982) Diurnal variation in exercise in angina pectoris. Br Heart J 48:156–160

Langner B, Lemmer B (1988) Circadian changes in the pharmacokinetics and cardiovascular effects of oral propranolol in healthy subjects. Eur J Clin Pharmacol 33:619–624

Lemmer B (1982) Pharmakologische Grundlagen der Therapie der koronaren Herzerkrankung mit Beta-Adrenozeptorenblockern. Herz 7:168–178

Lemmer B (1984) Chronopharmakologie – Tagesrhythmen und Arzneimittelwirkung, 2. Aufl. Wissenschaftliche Verlagsgesellschaft, Stuttgart

Lemmer B (1987) Chronopharmacology of cardiovascular medications. In: Kuemmerle HP, Hitzenberger G, Spitzy KH (eds) Klinische Pharmakologie, 4. Aufl. „ecomed", Landsberg München (12. Ergänzungslieferung, S 1–14)

Lemmer B (1988) Die Bedeutung der Chronopharmakologie für die medikamentöse Therapie. Verh Dtsch Ges Inn Med 94:420–430

Lemmer B (ed) (1989) Chronopharmacology – cellular and biochemical interactions. Marcel Dekker, New York Basel

Lemmer B, Bathe K (1982) Stereospecific and circadian-phase-dependent kinetic behaviour of d,l-, l- and d-propranolol in plasma, heart and brain of light-dark-synchronized rats. J Cardiovasc Pharmacol 4:635–644

Lemmer B, Labrecque G (1987) Chronopharmacology and chronotherapeutics: Definitions and concepts. Chronobiol Int 4:319–329

Lemmer B, Bathe K, Lang P-H, Neumann G, Winkler H (1983) Chronopharmacology of β-adrenoceptor blocking drugs: Pharmacokinetics and pharmacodynamic studies in rats. J Am Coll Toxicol 2:347–350

Lemmer B, Winkler H, Ohm T, Fink M (1985) Chronopharmacokinetics of beta-receptor blocking drugs of different lipophilicity (propranolol, metoprolol, sotalol, atenolol) in plasma and tissues after single and multiple dosing in the rat. Naunyn Schmiedebergs Arch Pharmacol 330:42–49

Lemmer B, Becker HJ, Renczes J, Scheidel B, Blume H (1986) Circadian-phase-dependency in the pharmacokinetics and cardiovascular effects of oral isosorbide dinitrate in man. Ann Rev Chronopharmacol 3:339–342

Lemmer B, Bärmeier H, Schmidt S, Lang P-H (1987) On the daily variation in the beta-receptor – adenylate cyclase – cAMP – phosphodiesterase system in rat forebrain. Chronobiol Int 4:469–475

Lemmer B, Behne S, Becker HJ (1989a) Chronopharmacology of oral nifedipine in healthy subjects. Eur J Clin Pharmacol [Suppl]36:A177

Lemmer B, Scheidel B, Stenzhorn G, Blume H, Lenhard G, Grether D, Renczes J, Becker HJ (1989b) Clinical chronopharmacology of oral nitrates. Z Kardiol [Suppl 3] 78:61–63

Marshall J (1977) Diurnal variation in occurrence of strokes. Stroke 8:230–231

Mitler MM, Hajdukovic RM, Shafor R, Hahn PM, Kripke DF (1987) When people die. Cause of death versus time of death. Am J Med 82:266–274

Muller JE, Stone PH, Turin ZG et al. (1985) The Milis study group: Circadian variation in the frequency of onset of acute myocardial infarction. N Engl J Med 313:1315–1322

Scheidel B, Blume H, Stenzhorn G, Lemmer B, Lenhard G (1987) Chronopharmacokinetics and chronohemodynamics of immediate-release isosorbide-5-mononitrate in man. Chronobiologia 14:233

Scheidel B, Blume H, Stenzhorn G, Lemmer B, Renczes J, Grether D, Becker HJ, Lenhard G (1988) Circadian variations in pharmacokinetics and hemodynamic effects of oral isosorbide dinitrate and isosorbide-5-mononitrate in man. (Symposium on Variability in Pharmacokinetics and Drug Response, Gothenburg, Sweden, Abstr 43)

Waters DD, Miller DD, Bouchard A, Bosch X, Theroux P (1984) Circadian variation in variant angina. Am J Cardiol 54:61–64

White WB, Smith VE, McCabe EJ, Meeran MK (1985) Effects of chronic nitrendipine on casual (office) and 24-h ambulatory blood pressure. Clin Pharmacol Ther 38:60–64

Yasue H, Omote S, Takizawa A, Nagao M, Miwa K, Tanaka S (1979) Circadian variation of exercise capacity in patients with Prinz-metal's variant angina: role of exercise-induced coronary arterial spasm. Circulation 59:938–948 rechnet, von denen 141 000 tödlich verlaufen sind. Interessanterweise verstarben innerhalb der ersten 4 h nach Schmerzbeginn bei den 25- bis 75jährigen Infarktpatienten bereits 26 % außerhalb des Krankenhauses. Diejenigen, die das Krankenhaus lebend erreichten, hatten eine deutlich bessere Überlebenswahrscheinlichkeit. Nach 4 Wochen waren insgesamt 59 % der Patienten verstorben, 34 % vor ärztlicher Behandlung.

Herzrhythmusstörungen wie Kammerflimmern und Asystolie dürften die Haupttodesursache der ersten 4 h sein, während die Herzinsuffizienz und kardiogener Schock an erster Stelle der Todesursachen der Spätphase stehen. Die ersten Stunden nach Schmerzbeginn im Rahmen des akuten Herzinfarkts haben in mehrfacher Hinsicht eine große Bedeutung; die prähospitale Sterblichkeit ist hoch, die diagnostischen Möglichkeiten sind durch den noch fehlenden pathologischen Anstieg myokardspezifischer Serumenzyme erschwert, und die therapeutischen Möglichkeiten, durch Thrombolyse die Infarktgröße zu begrenzen, sind auf diese Zeit begrenzt. Die Situation könnte durch Aufklärung der Bevölkerung, Laienreanimationsbemühungen und Notarztwagensysteme verbessert werden. Der schnelle Einsatz medizinischer Hilfe würde nicht nur die Ergebnisse thrombolytischer Therapie verbessern, sondern es könnten auch bei einem großen Teil der Patienten rechtzeitig lebensbedrohliche Herzrhythmusstörungen erkannt und behandelt werden.

Während kein sicheres Wissen über die frühe Entwicklung arteriosklerotischer Läsionen besteht, scheint es wenig Zweifel darüber zu geben, daß für die Progression der Arteriosklerose wandständige Thromben im Bereich einer Plaque mitverantwortlich sind. Der Grad der Durchmessereinengung bestimmt das weitere Schicksal bei Plaqueruptur. Bei hochgradigen Engen entwickeln sich fast regelmäßig verschließende Thromben, während Stenosen mit unbedeutender Durchmessereinengung oft nur zu intraintimalen Hämorrhagien führen (Falk 1983). Ohne daß der Triggermechanismus für die plötzliche Entwicklung eines thrombotischen Gefäßverschlusses bekannt ist, lassen sich bei seriellen Schnittuntersuchungen an Koronargefäßen von akut am Herzinfarkt verstorbenen Patienten Plaquerupturen in Form von Intimarissen, Dissektionen, Erosionen und Ulzerationen nachweisen. So finden sich im Querschnitt eines Koronargefäßes an der Verschlußstelle eine atheromatöse Plaque, intimale Fissuren, intraintimale Hämorrhagien und intraluminale Thromben.

Die Unterbrechung des koronaren Blutstromes führt zeitabhängig zum Untergang des Myokards (Hugenholtz et al. 1986). Die Nekrose beginnt in den subendokardialen Schichten wegen der größten Entfernung zu den epikardial gelegenen Koronargefäßen und der Auswirkung des hohen linksventrikulären Druckes und schreitet allmählich zu den subepikardialen Schichten fort (Reimer et al. 1977). Durch Wiederherstellung der Myokardperfusion wird die weitere Ausdehnung der Nekrose und eine zunehmende Verschlechterung der linksventrikulären Funktion verhindert. Wegen der Abhängigkeit der Morbidität und Letalität des Myokardinfarkts von der Ventrikelfunktion wird die Überlebenswahrscheinlichkeit vom Ausmaß der linksventrikulären Funktionsstörung bestimmt. Während nach 1 h weniger als 20 % des Myokards zugrundegegangen sind, dürften nach 3 h mehr als die Hälfte und nach 6 h

Therapie des akuten Myokardinfarktes mit Thrombolyse und PTCA

W. Rutsch und H. Schmutzler

Einleitung

Die Bedeutung der koronaren Herzkrankheit (KHK) in der Bundesrepublik Deutschland ergibt sich aus ihrem Stellenwert in der Todesursachenstatistik. Sie besetzt mit 21 % aller Todesfälle bei Männern und 16 % bei Frauen den ersten Platz. Obwohl 47 % der Männer und 69 % der Frauen älter als 75 Jahre sind, ist die KHK auch vor Erreichen des 75. Lebensjahrs bei 23 % der männlichen und 21 % der weiblichen Verstorbenen als Todesursache ausgewiesen. Nach Angaben des Augsburger Herzinfarktregisters (Löwel et al. 1985) erkrankten durchschnittlich 444 Männer und 138 Frauen je 100 000 der Bevölkerung im Verlauf des Kalenderjahres 1985 an einem Herzinfarkt. Die 28-Tage-Letalität betrug bei den Männern 54 % und bei den Frauen 66 %. Insgesamt verstarben 34 % der Erkrankten ohne in Krankenhausbehandlung gelangt zu sein (Abb. 1). Bei jedem 2. Verstorbenen war ein medizinischer Laie gegenwärtig und bei jedem 3. wurden notärztliche Reanimationsversuche unternommen.

Nach Angaben des Augsburger Herzinfarktregisters wurde für die Bundesrepublik Deutschland 1985 eine Rate akuter Herzinfarktfälle von 210 000 er-

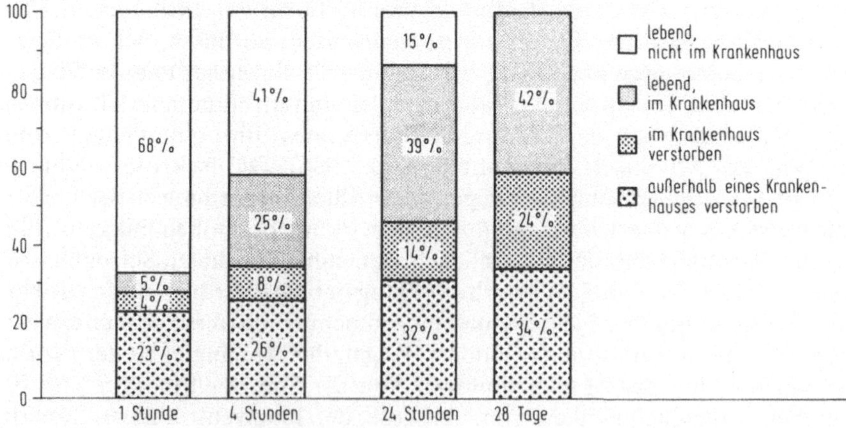

Abb. 1. Augsburger Infarktregister (MONICA-Projekt), Myokardinfarktsterblichkeit innerhalb der ersten 4 Wochen (zit. nach Dr. Loewel)

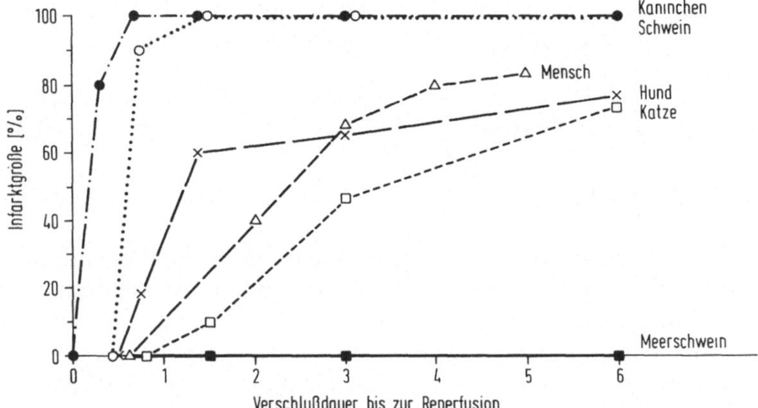

Abb. 2. Abhängigkeit der Infarktgröße von der Verschlußdauer. Verlauf der enzymatisch bestimmten Infarktgröße beim Menschen im Vergleich zu verschiedenen Tierspezies. (Nach: Hugenholtz PG, Schaper W (1986) Texas Heart Institute Journal 13:433–445)

nahezu das gesamte abhängige Myokard nekrotisch geworden sein (Abb. 2). Diese von Schaper und Hugenholtz erarbeitete zeitliche Beziehung der enzymatischen Infarktgrößenentwicklung wurde insbesondere durch die GISSI-Studie bestätigt, die einen klaren Zusammenhang des Zeitintervalls zwischen Schmerz- und Therapiebeginn und der Sterblichkeit nachweisen konnte (Gruppo italiano 1986). Im Gegensatz dazu stehen die Beobachtungen der ISIS-2-Studie (1988), bei der eine relative Senkung der Sterblichkeit durch Kurzzeitlyse mit Streptokinase noch bis zur 24. Stunde nach Schmerzbeginn nachgewiesen werden konnte. Wenn man davon ausgeht, daß innerhalb von 4–6 h das gesamte Myokard nekrotisch geworden ist, sofern vorher kein relevanter Kollateralkreislauf existierte, müssen zusätzliche Mechanismen diskutiert werden, die die Sterblichkeit beeinflussen können (Braunwald 1989).

Die Wiederherstellung des koronaren Blutflusses nach thrombotischem Gefäßverschluß kann nur durch Thrombolyse oder mechanische Intervention erreicht werden. Während die intrakoronare Infusion von Plasminogenaktivatoren auf die ersten Jahre dieses Therapieprinzips beschränkt blieb, wurde die systemische Thrombolyse in den letzten Jahren mit mechanischen Behandlungsverfahren, wie koronare Angioplastie, kombiniert. Gegenüber der Anwendung von Kathetertechniken, die die ständige Einsatzfähigkeit eines Katheterlabors voraussetzt, ist die systemische Thrombolyse technisch einfach, frühzeitig und ohne großen personellen Aufwand anwendbar.

Mehrere Studien aus den Jahren 1987, 1988 und 1989 haben die Bedeutung der koronaren Angioplastie im Stadium des akuten Infarkts relativiert. Übereinstimmend wurde berichtet, daß durch sofortige Angioplastie in Verbindung mit frühzeitig eingeleiteter thrombolytischer Behandlung bezüglich Infarktgröße, linksventrikulärer Funktion, Reststenose und Sterblichkeit kein Vorteil beobachtet werden konnte. Insofern ist in letzter Zeit zunehmend die Bedeutung der alleinigen systemischen Thrombolyse mit effektiven Plasminogenaktivatoren in den Vordergrund der Diskussion getreten.

Plasminogenaktivatoren

Streptokinase

Streptokinase findet bereits seit Jahrzehnten in der Behandlung thrombotischer und thrombembolischer Krankheitsbilder Anwendung. Die Mehrheit der frühen Studien mit intravenöser Applikation konnte keine signifikante Senkung der Sterblichkeit gegenüber Kontrollgruppen nachweisen. Da in den 60er und 70er Jahren der Zusammenhang zwischen Koronarthrombus und Infarktentwicklung nicht eindeutig geklärt war, wurde teilweise bis zur 48. Stunde nach Schmerzbeginn eine thrombolytische Behandlung durchgeführt. Yusuf fand bei Zusammenfassung von 7 prospektiv randomisierten Studien aus dieser Zeit eine Senkung der relativen Sterblichkeit nach systemischer Streptokinasetherapie von 15–20% (Yusuf et al. 1985).

Nach den ersten Arbeiten von Rentrop et al. (1980) mit intrakoronarer Infusion von Streptokinase folgten mehrere prospektiv-randomisierte Studien, die die hohe Effektivität dieser Therapieform belegen konnten, die Western-Washington-Studie (Kennedy et al. 1985) und die Untersuchung des Interuniversity Cardiology Institute in the Netherlands (ICIN; Simoons et al. 1985). Beide Studien waren durch eine konventionelle Behandlungsgruppe kontrolliert. Die Sterblichkeitsminderung war beträchtlich. Bei den in der Western-Washington-Studie erfaßten 250 Patienten lag die Sterblichkeit nach einem Monat hochsignifikant um 67% niedriger als in der Kontrollgruppe. Durch die relativ hohe Reinfarktrate ging ein Teil des Effekts auf die Sterblichkeit gegen Ende des ersten Jahres wieder verloren. Aber immerhin lag die Minderung der Sterblichkeit zu diesem Zeitpunkt noch bei 44%. In der ICIN-Studie mit 533 Patienten lag die Frühsterblichkeit nach einem Monat in der Streptokinasegruppe um 48% niedriger. Nur wenige Patienten verstarben innerhalb des nachfolgenden Jahres, der Unterschied betrug dann immerhin noch 44%.

Die unterschiedliche Reinfarktrate beider Untersuchungen macht deutlich, wie wichtig begleitende Behandlungsmaßnahmen wie Angioplastie und Koronarchirurgie nach erfolgreicher thrombolytischer Therapie sind. Die Analyse der linksventrikulären Funktionsdaten der ICIN-Studie zeigte eine signifikant bessere linksventrikuläre Funktion nach Thrombolyse im Vergleich zur konventionellen Vergleichstherapie. Die globale linksventrikuläre Ejektionsfraktion lag in der Thrombolysegruppe mit 53% hochsignifikant über der Kontrollgruppe mit nur 47%. Darüber hinaus war die regionale Myokardfunktion im Infarktareal in der Thrombolysegruppe signifikant besser, d.h. sowohl im Bereich der Hinter- als auch Vorderwand (Serruys et al. 1985). Bezüglich aller wesentlichen Parameter wie Rekanalisationsrate, regionale und globale Myokardfunktion sowie Sterblichkeit gilt gerade die holländische Studie als goldener Standard der thrombolytischen Therapie.

Erste überzeugende Ergebnisse mit intravenöser Kurzzeitlyse mit Streptokinase konnten mit der italienischen GISSI-Studie gewonnen werden (Gruppo italiano 1986), nachdem die von Schröder initiierte ISAM-Studie (1986) bei 1700 Patienten mit akutem Myokardinfarkt keinen Unterschied in der Frühsterblichkeit hatte nachweisen können. Bei 11712 randomisierten Patienten lag die Frühsterblichkeit nach 1,5 Mio. IE Streptokinase bei 11 %, gegenüber 13 % in der Kontrollgruppe. Dieser Unterschied war hochsignifikant. Bei einer Subgruppenanalyse wurde jedoch deutlich, daß sich der Unterschied in der Sterblichkeit auf die Patienten beschränkt, die innerhalb der ersten 6 h nach Schmerzbeginn behandelt wurden (Tabelle 1). Bei Patienten mit Hinter- oder Seitenwandinfarkt, bei Reinfarkten sowie bei Patienten mit Herzinsuffizienz wurde kein überzeugender Vorteil durch thrombolytische Therapie gesehen. In der Streptokinasegruppe traten signifikant häufiger Reinfarkte auf und die Häufigkeit zerebrovaskulärer Ereignisse war über beide Behandlungsgruppen mit 1 % relativ gleichmäßig verteilt. Verlaufsuntersuchungen über 1 Jahr bestätigten auch noch zu diesem späten Zeitpunkt den Vorteil frühzeitiger thrombolytischer Behandlung. Ein statistisch signifikanter Unterschied in der Sterblichkeit war bei der Einjahresmortalität nur bei den Patienten nachweisbar, die innerhalb von 6 h therapiert werden konnten. Im Gegensatz dazu stehen die Langzeitbeobachtungen der Western-Washington-Studie, bei der sich überraschenderweise ein Zusammenhang zwischen Spätletalität und Koronarstatus nach thrombolytischer Therapie ergab. Patienten mit vollständiger Reperfusion hatten die höchste Einjahresüberlebenswahrscheinlichkeit, während Patienten mit unzureichender oder erfolgloser thrombolytischer Behandlung deutlich höhere Sterblichkeitsraten aufwiesen. Hierin könnte eine Erklärung für die Ergebnisse der ISIS-2-Studie zu finden sein, daß auch nach der 6. Stunde und bereits kompletter Myokardnekrose der Koronarstatus einen zusätzlichen Einfluß auf die Sterblichkeit hat.

In der ISIS-2-Studie wurden 17187 Patienten mit infarktverdächtigen Thoraxschmerzen bis zu 24 h nach Krankheitsbeginn behandelt. Verglichen wurde eine systemische Kurzzeitlyse mit Streptokinase 1,5 Mio. IE über

Tabelle 1. GISSI-Studie, Letalität in Abhängigkeit vom Intervall zwischen Symptom- und Behandlungsbeginn

Zeitintervall (Stunden)	Todesfälle (%) (Anzahl/n)			
	Total	Konventionelle Therapie	Streptokinase-Therapie*	*p*
<3	10,6 (647/6094)	12,0 (369/3076)	9,2 (278/3016)	0,0005
>3–6	12,9 (471/3649)	14,1 (254/1800)	11,7 (217/1849)	0,03
>6–9	13,3 (180/1352)	14,1 (93/659)	12,6 (87/693)	ns
>9–12	14,6 (87/594)	13,6 (41/302)	15,8 (46/292)	ns
<1	11,8 (151/1277)	15,4 (99/642)	8,2 (52/635)	0,0001

* Streptokinase-Therapie i.v. 1,5 Mio. IE/60 min.
Nach: F. Rovelli et al., Lancet 1986, I: 397–401

60 min, eine Behandlung mit Acetylsalicylsäure 160 mg/Tag über einen Monat und die Kombination beider Pharmaka. Bis zur 24. Stunde nach Schmerzbeginn konnte mit Streptokinase oder Aspirin allein und insbesondere mit der Kombination eine hochsignifikante Senkung der Sterblichkeit gegenüber der jeweiligen Kontrollgruppe erreicht werden. Die Senkung der relativen Sterblichkeit für Streptokinase allein betrug 9,2 %, für Aspirin 9,4 % und für die Kombination beider Pharmaka 8 %. Die Kombination hatte einen additiven Effekt auf die Sterblichkeit, die bei Therapie innerhalb der ersten 4 h nach Schmerzbeginn um 53 % niedriger als in der Kontrollgruppe lag.

Das Protokoll der ISIS-2-Studie forderte als Einschlußkriterium nicht ein infarkttypisches EKG. Die retrospektive Analyse der Eingangs-EKG erbrachte daher infarkttypische Veränderungen bei nur etwas mehr als der Hälfte der Patienten. Die ISIS-2-Studie steht in der Subgruppenanalyse in einigen Punkten im Widerspruch zur GISSI-Studie. Ein so eindeutiger Zusammenhang zwischen Therapiebeginn und Sterblichkeitssenkung (Tabelle 2) bestand nicht, und es profitierten von der thrombolytischen Therapie auch die älteren Patienten sowie Patienten mit Hinterwand- und Reinfarkten. Bedeutende Nebenbefunde der ISIS-2-Studie waren neben dem hohen therapeutischen Stellenwert der Acetylsalicylsäure der nahezu vollständig fehlende therapeutische Effekt der Thrombolyse bei Patienten mit ST-Streckensenkungen oder ausgeprägter Herzinsuffizienz mit niedrigen systolischen Aortendrücken.

Obwohl die ISIS-2-Studie keine eindeutige diagnostische Trennung von ischämischem und hämorrhagischem Hirninfarkt zuließ, lag die Häufigkeit aller zerebrovaskulärer Ereignisse in der Placebogruppe höher als in der Gruppe mit Streptokinasetherapie. Aspirin hingegen senkte die Rate aller zerebrovaskulärer Ereignisse um 40 %. Am ausgeprägtesten war dieser Effekt wiederum in der Kombination von Streptokinase und Acetylsalicylsäure mit einer Senkung der Häufigkeit zerebrovaskulärer Ereignisse um 50 %. Obwohl es sich hier um eine Mischung ischämischer und hämorrhagischer zerebrovaskulärer Komplikationen handelte und die Komplikation der Thrombolyse eine meist tödliche Hirnblutung darstellt, traten zerebrovaskuläre Ereignisse insgesamt in der Kontrollgruppe häufiger auf. Wenn dies den tatsächlichen Gege-

Tabelle 2. ISIS-2-Studie, Zeitabhängigkeit der Senkung der relativen Sterblichkeit in den verschiedenen Behandlungsgruppen

Stunden	Streptokinase %	Aspirin %	Streptokinase + Aspirin %	Placebo %
0–1	8,1	9,6	6,2	13,0
2	7,6	8,4	6,5	11,5
3	8,5	8,8	6,6	12,4
4	8,5	9,2	6,1	12,2
5–12	10,1	10,1	9,8	12,6
13–24	8,7	8,7	7,4	10,9

benheiten entspräche, müßte gefolgert werden, daß zerebrovaskuläre Ereignisse nicht als Komplikation der Streptokinasebehandlung angesehen werden können.

Deutet man die zur Subgruppe mit Thoraxschmerzen und ST-Streckensenkungen gehörigen Patienten als solche im Zustand der instabilen Angina pectoris, so ergab sich in der ISIS-2-Studie für diese Gruppe kein Vorteil durch die thrombolytische Behandlung.

Urokinase

Im Vergleich zur Streptokinase bietet die Urokinase durch ihre fehlende Antigenität theoretische Vorteile. Die Substanz ist jedoch beim akuten Infarkt unzureichend untersucht, so daß exakte Empfehlungen zur Dosierung nicht gegeben werden können. Ihre Effektivität im Hinblick auf Rekanalisationsrate, linksventrikuläre Funktion und Sterblichkeit ist weitgehend unbekannt. In einer offenen Studie mit 50 Patienten, bei denen die thrombolytische Therapie innerhalb von 3 h nach Schmerzbeginn eingeleitet werden konnte, wurde bei einer Dosierung von 2 Mio. IE Urokinase als intravenöse Bolusinjektion eine Patencyrate 60 min nach Therapiebeginn von 60 % erreicht (Mathey et al. 1985). In einer Vergleichsstudie mit rt-PA wurden mit 2 Mio. IE Urokinase vergleichbare Rekanalisationsraten erreicht (Neuhaus et al. 1988).

Recombinant-tissue-type-Plasminogenaktivator (rt-PA)

Mit der gentechnologischen Verfügbarkeit von rt-PA wurde 1984 ein Plasminogenaktivator in die thrombolytische Therapie des akuten Infarktes eingeführt, der wesentliche theoretische Vorteile gegenüber Streptokinase und Urokinase besitzt; rt-PA hat als physiologischer Plasminogenaktivator besondere Eigenschaften, es besitzt eine erhöhte Bindungskapazität zu Fibrin und aktiviert Plasminogen an der Fibrinoberfläche einige hundertmal effektiver als im Plasma. Diese sog. Fibrinspezifität von rt-PA bewirkt auf der einen Seite eine höhere Thrombolyserate und hat andererseits einen geringer ausgeprägten Effekt auf das Gerinnungssystem des Blutes. Die Effektivität der Substanz wurde durch umfangreiche Studien der amerikanischen TIMI-Arbeitsgruppe und der Europäischen Kooperativen Studiengruppe (ECSG) untersucht. Die Ergebnisse der TIMI-Reperfusionsstudie (1985), einer Vergleichsuntersuchung von rt-PA und Streptokinase, zeigte deutliche Unterschiede in der Effektivität beider Substanzen bei Patienten mit akutem Infarkt, die innerhalb von 7 h nach Schmerzbeginn in die Untersuchung einbezogen werden konnten. Verglichen wurde die intravenöse Behandlung von 80 mg rt-PA über 3 h mit Streptokinase 1,5 Mio. IE über 1 h. Endpunkt dieser angiographischen Untersuchung, in die 219 Patienten einbezogen wurden, war der Koronargefäßstatus 90 min nach Therapiebeginn. In der rt-PA-Gruppe konnte bei 62 % der Patienten eine funktionell vollständige Reperfusion erreicht werden, in der

Streptokinasegruppe nur bei 31 % (Chesebro et al. 1987). Die Studie wurde vorzeitig abgebrochen, da der Unterschied hochsignifikant war. Über die Gesamtbeobachtungszeit von 90 min nach Therapiebeginn waren zu jedem Zeitpunkt durch tPA 2- bis 3mal soviele Gefäße wie unter Streptokinase rekanalisiert. Der Verlust von Plasmafibrinogen war in der rtPA-Gruppe deutlich geringer ausgeprägt als in der Streptokinasegruppe. Blutungskomplikationen und Reokklusion der Infarktarterie waren in beiden Gruppen vergleichbar häufig. Die Unterschiede zwischen rt-PA und Streptokinase wurden in dieser größten angiographischen Studie bei einer Subgruppenanalyse besonders deutlich. Bei Patienten mit komplettem Gefäßverschluß bei Lysebeginn, waren in der Streptokinasegruppe nur 26 % der Gefäße gegenüber 56 % in der rt-PA-Gruppe rekanalisiert.

Während die Effektivität von Streptokinase vom Thrombusalter abhängig war, konnte dies für rt-PA nicht beobachtet werden. Streptokinase war weniger gut wirksam, wenn der Schmerzbeginn mehr als 4 h zurücklag. Aber selbst bei den Patienten, die innerhalb von 4 h behandelt wurden, war während der ersten 40 min nach Lysebeginn die Rekanalisationsrate für rt-PA nahezu doppelt so hoch wie für Streptokinase (Abb. 3).

In eine Patency-Studie der Europäischen Kooperativen Studiengruppe wurden 129 Patienten 6 h nach Schmerzbeginn eingeschlossen und placebokontrolliert mit 0,75 mg rt-PA/kg KG behandelt (Verstraete et al. 1985 a). Bei strengen Ein- und Ausschlußkriterien lag die Patency-Rate für rt-PA bei 61 % gegenüber 21 % in der Placebogruppe. Bedeutende Komplikationen im Zusammenhang mit der Thrombolyse wurden nicht beobachtet, und es bildete sich nur ein sehr moderater Verlust von Fibrinogen aus. Nach 90 min waren im Blut immerhin noch 52 % des Ausgangswertes vorhanden.

In einer weiteren Studie der ECSG wurde die Effektivität von rt-PA und Streptokinase verglichen. 90 min nach Behandlung mit 0,75 mg rt-PA/kg KG über 90 min bzw. 1,5 Mio. IE Streptokinase über 60 min fand sich für rt-PA

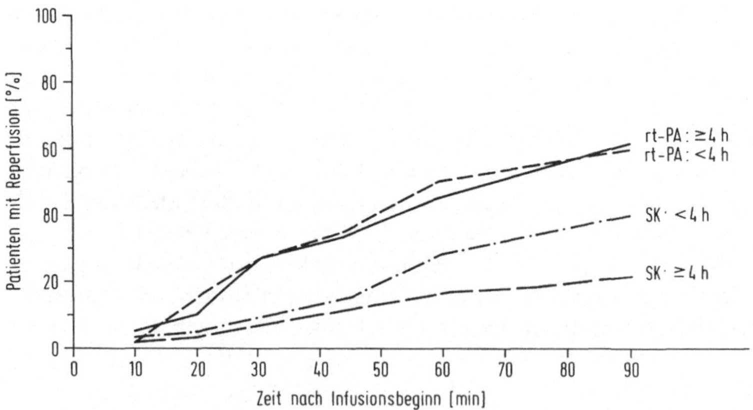

Abb. 3. TIMI-Studie, Wiedereröffnungsrate während der ersten 90 min nach Beginn der Thrombolyse, Vergleich rt-PA mit Streptokinase. (Nach: Chesebro JH (1987) Circulation 76:142–154)

eine Patencyrate von 70% gegenüber 55% für Streptokinase. Das Fibrinogen fiel in der rt-PA-Gruppe auf 61%, in der Streptokinasegruppe auf 12% ab (Verstraete et al. 1985b).

Restenosierung

Nach erfolgreicher thrombolytischer Therapie verbleibt in einem hohen Prozentsatz eine hochgradige Reststenose mit Plaqueruptur, intraintimalem Hämatom und Restthromben an der Oberfläche der Plaque. Reokklusion ist ein gewichtiges Problem, das bei erfolgreicher primärer thrombolytischer Therapie zum Reinfarkt führt, während bei kompletter Infarzierung die Reokklusion klinisch unbemerkt verläuft. Die Reokklusionsrate im Krankenhaus nach erfolgreicher Reperfusion mit intrakoronarer Streptokinase wird mit 5–29%, im Mittel 17%, angegeben. Angiographisch dokumentierte Wiederverschlußraten nach systemischer Thrombolyse mit Streptokinase liegen in einem ähnlichen Bereich (Rentrop et al. 1981; Ganz et al. 1983; Serruys et al. 1983; Neuhaus et al. 1983; Rogers et al. 1983; Schröder et al. 1983; Spann et al. 1984; Harrison et al. 1984). Wegen der kurzen fibrinolytischen Halbwertszeit von rt-PA und seines geringen Einflusses auf Gerinnungsparameter war für diesen Plasminogenaktivator eine höhere Reokklusionsrate vermutet worden. Im Gegensatz zum Kurzzeitlysekonzept mit Streptokinase von Schröder et al. (1983) mit einer Infusionszeit von 60 min war daher die thrombolytische Behandlung mit rt-PA anfänglich auf 3 h mit einem höher dosierten Anteil in der 1. Stunde festgelegt worden. Die Untersuchung von Gold et al. (1986) über die Reokklusionsrate nach rt-PA erbrachte eine hohe Reokklusionsrate. In einer Untersuchung an einer kleinen Gruppe von 29 Patienten beobachtete Gold nach erfolgreicher Rekanalisation unter Fortführung der Therapie mit Heparin eine Wiederverschlußrate von 45% innerhalb von 1 h nach Ende der rt-PA-Infusion. Die 2. Gruppe von Patienten erhielt nach vollständiger Reperfusion eine Anschlußbehandlung mit 10 mg rt-PA/h über 4 h. Kein Patient erlitt eine Reokklusion während der Hospitalzeit. Zwischen dem Grad der Reststenose, dem Plasma-rt-PA-Spiegel und der Wiederverschlußrate bestand eine enge Korrelation.

Die Frage der Reokklusionshäufigkeit nach rt-PA Therapie wurde in einer großen prospektiv randomisierten Studie der ECSG untersucht (Verstraete et al. 1987). 123 Patienten erhielten rt-PA innerhalb von 4 h nach Schmerzbeginn. Nach 90 min wurde die Durchgängigkeit des Infarktgefäßes angiographisch bewertet. 66% der Patienten mit erfolgreicher Thrombolyse wurden zu Placebo oder einer weiteren rt-PA-Infusion mit 30 mg über 6 h randomisiert. Ein zweites Angiogramm 6–24 h nach Randomisierung zeigte keinen signifikanten Unterschied in der Wiederverschlußrate, 6% in der aktiven gegenüber 8% in der Placebogruppe.

Anisoylierter Plasminogen-Streptokinase-Aktivatorkomplex (APSAC)

APSAC ist auf der aktiven Seite des Plaminogenmoleküls durch eine Anisoylgruppe chemisch geschützt, so daß nach systemischer Injektion die unspezifische Aktivierung und Fibrinogenolyse vermieden werden kann. Andererseits scheint die Fähigkeit, sich an Fibrin zu binden, verstärkt zu sein. APSAC wird semiselektiv am Thrombus durch Hydrolyse aktiviert. Die fibrinolytische Aktivität ist bei einer Plasmahalbwertszeit von 120 min deutlich verzögert. Experimentielle Untersuchungen haben eine höhere Thrombusbindung und -lyse für APSAC ergeben und weniger ausgeprägte Blutdruckabfälle als nach Streptokinase. Bei etwa gleicher Nebenwirkungsrate scheint die thrombolytische Aktivität höher zu sein als die von Streptokinase.

Es gibt eine Vielzahl von Reperfusionsstudien mit APSAC. Die meist kleinen Probandenzahlen lassen jedoch keine sichere Aussage zu. In einer koronarangiographischen Studie von Anderson et al. (1988) wurde APSAC (30 IE/2–4 min) mit einer intrakoronaren Streptokinaseinfusion (160000 IE/60 min) verglichen. Für APSAC wurde eine Reperfusionsrate von 51 % und für Streptokinase von 60 % beobachtet. Der Unterschied war nicht signifikant. Der Plasmafibrinogenspiegel fiel in der 90. Minute unter APSAC sehr viel deutlicher ab, und es wurden mehr Blutungskomplikationen beobachtet. Die Reokklusionsrate war in beiden Gruppen identisch, ebenso der enzymatische Kurvenverlauf. Während die mittlere Reperfusionszeit für APSAC bei 43 min lag, war sie mit 31 min für die intrakoronare Streptokinasetherapie deutlich kürzer. Der Erfolg von APSAC war vom Zeitintervall zwischen Schmerz- und Therapiebeginn abhängig. 60 % der Infarktgefäße waren bei Behandlungsbeginn innerhalb von 4 h, jedoch nur 33 % bei längerem Intervall rekanalisiert. Bei Patienten mit komplettem Gefäßverschluß war APSAC ebenfalls der Streptokinase unterlegen: 43 % gegenüber 54 % Rekanalisationsrate.

Obwohl APSAC mit einem etablierten Thrombolysebehandlungsschema verglichen werden sollte, muß kritischerweise bemerkt werden, daß bei der Mehrheit der Studien mit intrakoronarer Streptokinaseinfusion wesentlich höhere Dosen als 160000 IE verwendet wurden. Andererseits wurden die Rekanalisationsraten beider Pharmaka zu unterschiedlichen Zeiten ermittelt, für Streptokinase nach 60 min – am Ende der intrakoronaren Infusion – und für APSAC nach 90 min. Wäre für Streptokinase eine höhere Dosierung und das gleiche Zeitintervall gewählt worden, wäre der Unterschied zu APSAC sehr viel deutlicher ausgefallen. Daß unterschiedlich effektive Dosierungen Anwendung fanden, wird auch am Verhalten der Gerinnungsparameter deutlich: stärkerer Abfall des Fibrinogens und mehr Blutungskomplikationen in der APSAC-Gruppe.

Vorläufige Ergebnisse einer Sterblichkeitsstudie mit APSAC (APSAC Intervention Mortality Studie (AIMS Trial Study Group 1988) ergaben bei der 30-Tage-Letalität eine Senkung der Sterblichkeit um 47 % gegenüber einer placebobehandelten Kontrollgruppe. Die Wirksamkeit von APSAC war unabhängig vom Lebensalter, der Infarktlokalisation und dem Behandlungsbeginn innerhalb der ersten 6 h.

Zusammenfassender Vergleich der Plasminogenaktivatoren

Streptokinase und Urokinase sind unspezifische, systemisch wirkende Plasminogenaktivatoren, die durch die Bildung von Plasmin im Plasma Thromben lysieren. Neben der Spaltung von Fibrin werden Gerinnungsfaktoren und Fibrinogen verbraucht, wodurch ein anhaltender Hämostasedefekt entsteht. Selbst die Kurzzeitlyse mit Streptokinase bewirkt bereits nach 90 min einen ausgeprägten systemischen Gerinnungsdefekt. Die Effektivität von Streptokinase, koronare Thromben zu lysieren, hat sich in den angiographisch kontrollierten Studien als gering erwiesen. Streptokinase muß infundiert werden, da ansonsten der Plasminanstieg einen gewichtigen Blutdruckabfall auslösen würde (Tabelle 3).

Für Urokinase, die in der Bundesrepublik Deutschland für die systemische Behandlung des akuten Infarktes nicht zugelassen ist, fehlen überzeugende Dosis-Wirkungs-Beziehungen, obwohl sie durch ihre fehlende Antigenität gegenüber Streptokinase theoretische Vorteile besitzt. Die Bewertung von APSAC ist aus der vorliegenden Literatur nur mit Einschränkung möglich. APSAC besitzt gegenüber Streptokinase keine Vorteile im Hinblick auf die Nebenwirkungsrate. Mit einer veränderten Kinetik kann noch keine Fibrinspezifität im eigentlichen Sinne begründet werden. Die koronarangiographisch kontrollierte Rekanalisationsrate von 51 % liegt höher als für Streptokinase dokumentiert werden konnte, bleibt jedoch deutlich unter den Rekanalisationsraten von rt-PA. Andererseits konnte mit der AIMS-Studie eine hohe Senkung der relativen Sterblichkeit mit APSAC nachgewiesen werden.

Tabelle 3. Vergleich der Eigenschaften verfügbarer Plasminogenaktivatoren

	Streptokinase	Urokinase	Acylierter Streptokinase-Plasminogen-Komplex	Gewebe-Plasminogen-Aktivator (t-PA)
Syst. Aktivierung des fibrinol. Systems mit Fibrinogenolyse	+++	+++	++	–
Fibrinspezifität	–	–	(–)	++
Antigenität/Antikörper	ja	nein	ja	nein
Blutungsrisiko	ja	ja	ja	geringer
Andere Nebenwirkungen	ja	nein	ja	nein
Bolus oder Infusion	Infusion	Infusion	Bolus-Injektion	Infusion
Halbwertszeit	40–60 min	10–15 min	1,7 h	3–5 min
Rekanalisationsrate	31%	nicht bestimmt	52–64%	71–76%

PTCA beim akuten Myokardinfarkt

Einleitung

Nach erfolgreicher Thrombolyse verbleibt bei etwa 75 % aller Patienten eine Reststenose von mehr als 75 % (Rutsch et al. 1982). Bei etwa zeitgleicher Entwicklung der thrombolytischen Behandlung und koronaren Angioplastie war es naheliegend, im Rahmen der intrakoronaren Applikation von Plasminogenaktivatoren die Dilatation der verbliebenen Koronarstenose anzuschließen. 1982 berichteten erstmals Meyer et al., daß mit PTCA nach erfolgreicher Thrombolyse die Prognose verbessert werden kann. In eienr prospektiv randomisierten Untersuchung fand Meyer seltener Reokklusionen, Reinfarkte und niedrigere Sterblichkeitsraten in der Gruppe mit Angioplastie.

Drei weitere Untersuchungen der jüngeren Zeit beschäftigten sich mit dem Stellenwert der PTCA im Rahmen des akuten Myokardinfarkts, die Untersuchung der TAMI-, ECSG- und TIMI-Arbeitsgruppe. In allen 3 Untersuchungen wurde rt-PA als Plasminogenaktivator bei unterschiedlichen Fragestellungen verwendet.

Offensichtlich ist die Kombination frühzeitiger Thrombolyse mit sofortiger Angiographie und PTCA aufgrund der Gegebenheiten im Gefäßverschlußbereich – Thrombusreste, Plaqueruptur, intraintimales Hämatom und Unregelmäßigkeiten der Oberfläche – durch eine hohe Rate lokaler Komplikationen problematisch. Die gleichzeitige Anwendung von Plasminogenaktivatoren und mechanischer Intervention mit gewichtiger zusätzlicher Intimaläsion führt häufig zu instabiler Angina pectoris und Reinfarkt oder Reokklusion. Die alleinige Anwendung von PTCA ohne Thrombolyse scheint mit weniger Komplikationen behaftet zu sein.

TAMI-Studie

In der TAMI-Studie (Topol et al. 1988) wurde der Einfluß sofortiger gegenüber verzögerter PTCA auf die regionale und globale linksventrikuläre Myokardfunktion untersucht. 90 min nach Beginn der Thrombolyse waren 75 % der Infarktgefäße rekanalisiert. Die Sterblichkeit bei den Thrombolyseversagern war mit 10,4 % gegenüber 7 % der Gesamtgruppe deutlich höher. Die erfolgreich behandelten Infarktpatienten wurden in die Studie einbezogen, sofern die Koronargefäßanatomie für eine Behandlung mit PTCA geeignet erschien. Patienten mit linkskoronarer Hauptstammstenose, diffuser Koronarsklerose und Dreigefäßerkrankung wurden ausgeschlossen und hatten mit 11 % wiederum eine hohe Sterblichkeit. Die Gruppe der erfolgreich lysierten und für eine PTCA geeigneten Patienten hatte mit 2,5 % eine ausgesprochen niedrige Sterblichkeit; 98 Patienten wurden zur späten PTCA-Gruppe am 7. Tag randomisiert (Abb. 4).

Aus verschiedenen Gründen wurde die PTCA zum späten Zeitpunkt jedoch nur bei 35 Patienten durchgeführt, bei dem Rest war eine vorzeitige

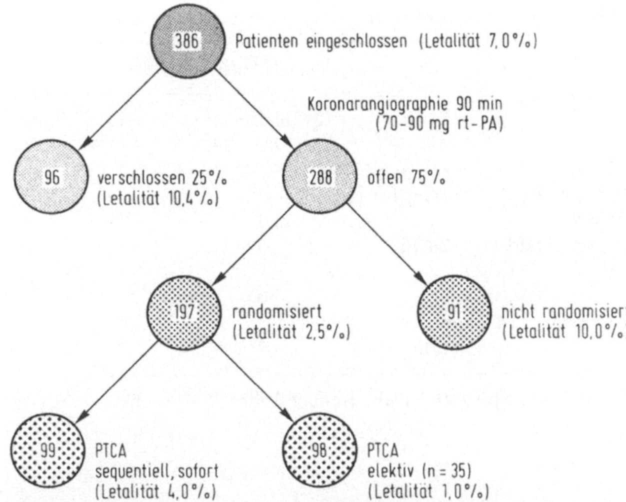

Abb. 4. TAMI-Studie, Erfolg der Thrombolyse mit rt-PA, Vergleich der Sterblichkeit in den Gruppen Thrombolyseversager, für PTCA ungeeignete Gefäßanatomie sowie frühe und späte PTCA

PTCA oder notfallmäßige Bypass-Chirurgie erforderlich, oder es zeigten sich bei der Kontrollkoronarangiographie Reststenosen, die eine PTCA nicht mehr erforderlich machten. Insofern ist ein Gruppenvergleich problematisch. Bezüglich der primären Endgröße, globale und regionale linksventrikuläre Funktion, ergaben sich keine Unterschiede. Jedoch war die Sterblichkeit in der Gruppe mit früher PTCA deutlich höher, als in der spät behandelten Gruppe. Es muß jedoch betont werden, daß die TAMI-Studie keine Letalitätsstudie war, der Probandenumfang war viel zu gering. Die Schlußfolgerung der Untersuchung, daß bei Patienten mit erfolgreicher Thrombolyse mit rt-PA und einer für PTCA geeigneten Anatomie, die frühzeitige PTCA keine zusätzlichen Vorteile bietet, stimmt jedoch mit den Schlußfolgerungen der anderen großen Untersuchungen überein.

ECSG-Studie

Das Konzept der PTCA-Studie der Europäischen Kooperativen Studiengruppe (Simoons et al. 1988) war der Vergleich einer maximalen Therapiestrategie – Thrombolyse, Koronarangiographie und PTCA – mit alleiniger thrombolytischer Behandlung. Nach Thrombolyse mit rt-PA und vor Koronarangiographie erfolgte die Randomisierung. Auch in dieser Studie war nicht die Sterblichkeit primärer Endpunkt, sondern die enzymatische Infarktgröße, globale linksventrikuläre Funktion und der klinische Verlauf. Wegen deutlich höherer Sterblichkeit in der invasiven Gruppe wurde die Untersuchung vorzei-

Tabelle 4. PTCA-Studie der ECSG, klinischer Verlauf

	Häufigkeit (%)	
	t-PA $n=184$	t-PA/PTCA $n=183$
Hypotonie (<90 mmHg)	9	31
Schock	3	6
Ischämierezidiv (<24 h)	3	17
(24 h–14 Tage)	11	13
(24 h–3 Monate)	30	28
Todesfälle 14 Tage	3	7
3 Monate	3	8

Nach: M.L. Simoons et al., Lancet 1988, I: 197–202

tig abgebrochen (Tabelle 4). Unterschiede beider Gruppen im Hinblick auf die primären Endgrößen ergaben sich nicht. In der PTCA-Gruppe wurde in den ersten 24 h eine hohe Rate instabiler Angina pectoris beobachtet. Nach 14 Tagen war die Sterblichkeit in der Thrombolysegruppe mit 3 % bei 367 eingeschlossenen Patienten deutlich niedriger als mit 7 % in der invasiven Gruppe. Da kein zusätzlicher Vorteil durch PTCA nach Thrombolyse erkennbar wurde, lautete die Empfehlung auch dieser Studie, auf Angiographie und PTCA im akuten Infarktstadium zu verzichten. Der methodische Nachteil der ECSG-Untersuchung war in der Tatsache begründet, daß die Randomisierung vor Angiographie erfolgte und damit Patienten der PTCA Gruppe zugeordnet wurden, die eine für PTCA ungeeignete Koronargefäßanatomie aufwiesen.

TIMI II

In die Untersuchung der TIMI Study Group (1989) wurden 3262 Patienten eingeschlossen die mit rt-PA intravenös innerhalb der ersten 4 h nach Schmerzbeginn behandelt wurden. Die Hälfte der Patienten wurde zu Koronarangiographie mit PTCA 18–48 h nach Beginn der thrombolytischen Behandlung randomisiert. Die PTCA wurde nur bei Patienten mit geeigneter Koronargefäßanatomie durchgeführt. In der Kontrollgruppe wurden die Patienten einer konservativen Behandlungsstrategie unterzogen, wobei Koronarangiographie und PTCA nur bei spontan auftretender oder belastungsinduzierter Myokardischämie indiziert war. In der Gruppe mit verzögerter PTCA wurde der Eingriff bei nur 57 %, im Gegensatz zu 13 % bei konservativer Behandlungsstrategie durchgeführt. Die primäre Endgröße der Untersuchung, Reinfarkt oder Tod innerhalb von 42 Tagen, betrug für die Gruppe mit invasiver Behandlungsstrategie 10,9 % gegenüber 9,7 % bei konservativem Vorgehen. Der Unterschied war statistisch nicht signifikant. Ein Unterschied der Ejektionsfraktion beider Gruppen ließ sich nicht nachweisen. Schlußfolgerung der Untersuchung war, daß die prophylaktische PTCA nach 18–48 h gegenüber einer abwartenden Haltung keinen Vorteil im Hinblick auf Senkung der Sterblichkeit, Limitierung der Infarktgröße, Verbesserung der Ventrikelfunktion und Reinfarktrate bietet. Es wurde empfohlen, Angiographie und PTCA den Patienten vorzubehalten, bei denen im Postinfarktverlauf erneut Beschwerden auftreten.

Schlußfolgerungen

Die systemische Thrombolyse bei akutem Myokardinfarkt ist zu einer allgemein akzeptierten Therapieform geworden, sofern wegen des Blutungsrisikos Ein- und Ausschlußkriterien eingehalten werden. Die Überlebenswahrscheinlichkeit wird von der Infarktgröße und der Ventrikelfunktion bestimmt, über die in den ersten 6 h entschieden wird. Offensichtlich scheint es jedoch nicht bedeutungslos zu sein, ob trotz kompletter Infarzierung das versorgende Gefäß offen oder verschlossen ist. Vieles spricht dafür, daß über die Bewahrung der Ventrikelfunktion hinaus für die Letalität auch ein offenes Infarktgefäß von Bedeutung ist. Und dies ist nicht von einer frühen Intervention abhängig. So kann die Prognose des Infarktpatienten durch 2 unterschiedliche Mechanismen beeinflußt werden, wobei ohne Zweifel der Begrenzung der Infarktgröße in der Frühphase die größte Bedeutung zukommt. Offensichtlich lohnt sich jedoch auch noch eine Behandlung mit Plasminogenaktivatoren jenseits der ersten 6 h, wie dies in der ISIS-2-Studie gezeigt werden konnte und aus der Langzeitbeobachtung der Western-Washington-Studie mit intrakoronarer Streptokinaseeinfusion vermutet werden konnte.

Das Konzept früher systemischer Thrombolyse in Kombination mit sofortiger oder um Stunden verzögerter PTCA hat sich nicht bewährt. Alle prinzipiellen Forderungen an diese Behandlung haben sich, bis auf die hohe primäre Rekanalisationsrate, nicht erfüllt. Die primäre PTCA ohne Thrombolyse hat jedoch dort, wo sie problemlos wegen eines einsatzfähigen Katheterlabors anwendbar ist, weiterhin einen gewichtigen Stellenwert. Die Komplikationsraten sind niedrig, es lassen sich sehr gute Früh- und Spätergebnisse erzielen und die Behandlung kommt für die große Gruppe von Patienten in Betracht, bei denen eine Kontraindikation gegen eine Thrombolyse besteht. Immerhin sind in den großen Studien bis zu 30 % der Patienten wegen Kontraindikationen ausgeschlossen worden. Die technische Weiterentwicklung der Angioplastie verspricht auch bei dieser Indikation in Zukunft gewichtige Fortschritte.

Als allgemeine Empfehlung kann gelten: Fibrinolyse so früh wie möglich, jedoch auch ausnahmsweise bis zur 24. Stunde, ggf. auch schon prästationär mit dem wirksamsten Plasminogenaktivator, um dann eine abwartende Haltung einzunehmen. Koronarangiographie und weitere Maßnahmen sollten dann von der klinischen Entwicklung abhängig gemacht werden. Bei erneuter Entwicklung von Beschwerden oder positiven Belastungstests vor Entlassung sollte unbedingt eine angiographische Klärung erfolgen, um Informationen über weitergehende Behandlungsmöglichkeiten zu gewinnen.

Wünschenswert für die weitere Entwicklung wäre eine umfassendere Aufklärung der Bevölkerung, um mehr als durchschnittlich nur 10–20% der Infarktpatienten einer thrombolytischen Behandlung zuführen zu können, sowie die Organisation prästationärer Therapieeinleitung mit entsprechend ausgerüsteten Notarztwagensystemen und eine breitere Anwendung effektiverer und nebenwirkungsfreier Plasminogenaktivatoren. Mit welchem der verfügbaren Pharmaka die Sterblichkeit am effektivsten positiv beeinflußt werden kann, wird mit der GISSI-2- und der ISIS-3-Studie in absehbarer Zukunft entschieden sein.

Literatur

AIMS Trial Study Group (1988) Effect of intavenous APSAC on mortality after acute myocardial infarction: preliminary report of a placebo-controlled clinical trial. Lancet II:441–446

Anderson JL, Rothbard RL, Hackworthy RA et al. (1988) Multicenter reperfusion trial of intravenous anisoylated plasminogen streptokinase activator complex (APSAC) in acute myocardial infarction: controlled comparison with intracoronary streptokinase. J Am Coll Cardiol 11:1153–1163

Braunwald E (1989) Myocardial reperfusion, limitation of infarct size, reduction of left ventricular dysfunction, and improved survival. Should the paradigm be expanded? Circulation 79/2:441–444

Chesebro JH, Knatterud G, Roberts R et al. (1987) Thrombolysis in myocardial infarction (TIMI) trial, phase I: a comparison between intravenous tissue plasminogen activator and intravenous streptokinse. Clinical findings through hospital discharge. Circulation 76/1:142–154

Falk E (1983) Plaque rupture with severe pre-existing stenosis precipitating coronary thrombosis. Br Heart J 50:127–134

Ganz W, Geft I, Maddahi J, Berman D, Charuzi Y, Shah PK, Swan HJC (1983) Nonsurgical reperfusion in evolving myocardial infarction. J Am Coll Cardiol 1:1247–1250

GISSI: Gruppo italiano per lo studio della streptochinasi nell infarto miocardico (1986) Effectiveness of intravenous thrombolytic treatment in acute myocardial infarction. Lancet 8478:397–401

Gold HK, Leinbach RC, Garabedian HD, Yasuda T, Johns JA, Grossbard EB, Palacios I, Collen D (1986) Acute coronary reocclusion after thrombolysis with recombinant human tissue-type plasminogen activator: prevention by a maintenance infusion. Circulation 73,2:347–352

Harrison DG, Ferguson DW, Collins SM, Skorton DJ, Ericksen EE, Koschos JM, Marcus ML, White CW (1984) Rethrombosis after reperfusion with streptokinase: importance of geometry of residual lesions. Circulation 69:991–996

Hugenholtz PG, Serruys PW, Simoons ML, Lubsen J (196) Thrombolytic therapy for acute coronary obstruction: Status in 1986. Texas Heart Inst J 13:433–445

ISAM Study Group (1986) A prospective trial of intravenous streptokinase in acute myocardial infarction (I.S.A.M.). Mortality, morbidity and infarct size at 21 days. N Engl J Med 314:1465–1471

ISIS-2 (Second International Study of Infarct Survival) Collaborative Group (1988) Randomised trial of intravenous streptokinase, oral aspirin, both, or neither among 17187 cases of suspected acute myocardial infarction: ISIS-2. Lancet 8607:349–360

Kennedy JW, Ritchie JL, Davis KB, Stadius ML, Maynard C, Fritz JK (1985) The Western Washington randomized trial of intracoronary streptokinase in acute myocardial infarction. N Engl J Med 309:1477–1482

Löwel H, Lewis M, Keil U, Koenig W, Hormann A, Bolte HD, Gostomzyk J (1988) Zur Herzinfarktsituation in einer süddeutschen Bevölkerung: Ergebnisse des Augsburger Herzinfarktregisters 1985. Z Kardiol 77:481–489

Mathey DG, Schofer J, Sheehan FH, Becher H, Tilsner V, Dodge HT (1985) Intravenous urokinase in acute myocardial infarction. Am J Cardiol 55:878–882

Meyer J, Merx W, Dörr R, Lambertz H, Bethge C, Effert S (1982) Successful treatment of acute myocardial infarction shock by combined percutaneous transluminal coronary recanalization (PTCR) and percutaneous transluminal coronary angioplasty (PTCA). Am Heart J 103:132–136

Neuhaus KL, Tebbe U, Sauer G, Kreuzer H, Kostering H (1983) High dose intravenous streptokinase in acute myocardial infarction. Clin Cardiol 6:426–430

Neuhaus KL, Tebbe U, Gottwik M et al. (1988) Intravenous recombinant tissue plasminogen activator (rt-PA) and urokinase in acute myocardial infarction: Results of the German Activator Urokinase Study (GAUS). J Am Coll Cardiol 12:581–587

Reimer KA, Lowe JE, Rasmussen MM, Jennings W (1977) The wavefront phenomen of ischemic cell death. Circulation 56:786–794

Rentrop P, Blanke H, Köstering K, Karsch KR (1980) Intrakoronare Streptokinase Applikation beim akuten Infarkt und instabiler Angina pectoris. Dtsch Med Wochenschr 105:221–228

Rentrop P, Blanke H, Karsch KR et al. (1981) Changes in left ventricular function after intracoronary streptokinase infusion in clinically evolving myocardial infarction. Am Heart J 102:1188–1192

Rogers WJ, Mantle JA, Hood WP, Baxley WA, Whitlow PL, Reevers RC, Soto B (1983) Prospective randomized trial of intravenous and intracoronary streptokinase in acute myocardial infarction. Circulation 68:1051–1054

Rutsch W, Schartl M, Mathey D et al. (1982) Perkutane, transluminale koronare Rekanalisation: Methodik, Ergebnisse und Komplikationen. Z Kardiol 71:7–12

Schröder R, Biamino G, Leitner ER von, Linderer T, Bruggemann T, Heitz J, Vohringer HF, Wegscheider K (1983) Intravenous short-term infusion of streptokinase in acute myocardial infarction. Circulation 76:536–541

Serruys PW, Wijns W, Brand M van den et al. (1983) Is transluminal coronary angioplasty mandatory after successful thrombolysis? Quantitative coronary angiography study. Br Heart J 50:257–261

Serruys PW, Simoons ML, Suryapranata H et al. (1985) Preservation of global and regional left ventricular function after early thrombolysis in acute myocardial infarction. J Am Coll Cardiol 7:729–742

Simoons ML, Serruys PW, Brand M et al. (1985) Improved survival after early thrombolysis in acute myocardial infarction. A randomized trial by the Interuniversity Cardiology Institute in the Netherlands. Lancet II:578–581

Simoons ML, Arnold AER, Betriu A et al. (1988) Thrombolysis with tissue plasminogen activator in acute moycardial infarction: no additional benefit from immediate percutaneous coronary angioplasty. Lancet 8579:197–202

Spann JF, Sherry S, Carabello BA et al. (1984) Coronary thrombolysis by intravenous streptokinase in acute myocardial infarction: acute and follow-up studies. Am J Cardiol 53:655–660

TIMI: The thrombolysis in myocardial infarction trial (1985) Phase I findings. N Engl J Med 312:932–936

The TIMI Study Group (1989) Comparison of invasive and conservative strategies after treatment with intravenous tissue plasminogen activator in acute myocardial infarction. Results of the thrombolysis in myocardial infarction (TIMI) phase II trial. N. Engl J Med 320:618–627

Topol EJ, Califf RM, George BS et al. (1988) A randomized trial of immediate versus delayed elective angioplasty after intravenous tissue plasminogen activator in acute myocardial infarction. N Engl J Med 317:581–588

Verstraete M, Bleifeld W, Brower RW et al. (1985a) Double-blind randomised trial of intravenous tissue-type plasminogen activator versus placebo in acute myocardial infarction. Lancet II:965–969

Verstraete M, Bernard R, Bory M et al. (1985b) Randomised trial of intravenous recombinant tissue-type plasminogen activator versus intravenous streptokinase in acute myocardial infarction. Lancet I:842–847

Verstraete M, Arnold AER, Brower RW et al. (1987) Acute coronary thrombolysis with recombinant human tissue-type plasminogen activator: initial patency and influence of maintained infusion on reocclusion rate. Am J Cardiol 60:231–237

Yusuf S, Collins R, Peto R, Furberg C, Stampfer MJ, Goldhaber SZ, Hennekes CH (1985) Intravenous and intracoronary fibrinolytic therapy in acute myocardial infarction: overview of results on mortality, reinfarction and side effects from 33 randomized controlled trials. Eur Heart J 6:556–585

Perkutane Transluminale Koronarangioplastie. Neuere Entwicklungen – Komplikationen und Empfehlungen

E. Fleck, E. Frantz, J. Krülls-Münch, U. Sauer und H. Oswald

PTCA und neue Verfahren

Seit ihrer Einführung in die Klinik ist die PTCA als Methode zur Revaskularisation stenosierter Koronargefäße generell akzeptiert worden. Sie trat neben die schon zuvor entwickelte Bypass-Chirurgie und wird derzeit an mehr Patienten eingesetzt als die operative Methode. In der Bundesrepublik Deutschland wurden 1988 durchgeführt (nach Gleichmann u. Mannebach 1989):

PTCA an 16923 Patienten,
ACVB-Operation an 21363 Patienten.

1988 wurden allein in den USA mehr als 200000 Koronarangioplastien durchgeführt (Lange u. Hillis 1989). Abbildung 1 zeigt die vergleichenden Zahlen aus dem Jahre 1988 für verschiedene Länder.

Die Hauptvorteile der PTCA liegen in der schnelleren Einsetzbarkeit, in der geringeren Invasivität und – damit verbunden – im erheblich geringeren Aufwand der Behandlung. Dem steht die Restenosierungsrate gegenüber, die

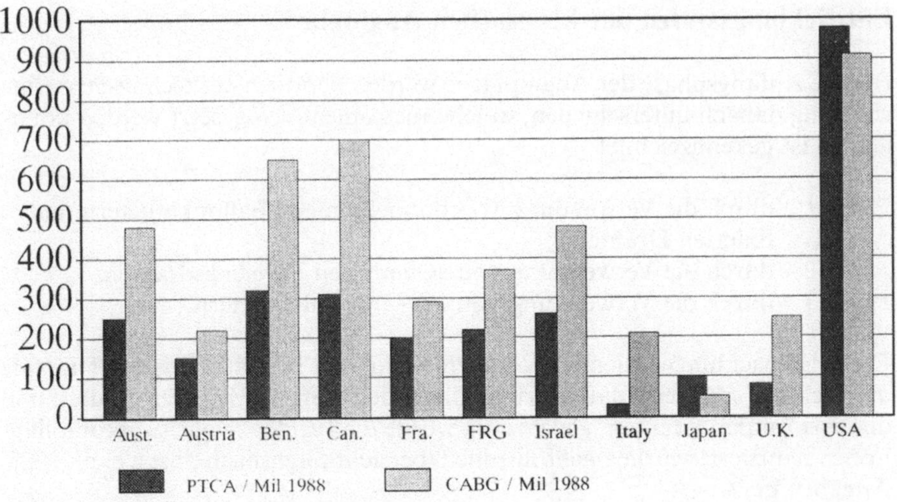

Abb. 1. Zahl der PTCA bzw. ACVB-Operationen (CABG) pro eine Million Einwohner in verschiedenen Ländern 1988

in verschiedenen untersuchten Gruppen unterschiedlich hoch, aber im Schnitt unverändert um 30% liegt (Holmes et al. 1984; Liu et al. 1989). Die Einführung quantitativer Auswertungsprogramme hat zur Aufdeckung einer eher noch ungünstigeren Restenosierungsrate geführt (Fleck et al. 1985).

Nicht zuletzt wegen der hohen Restenosierungsquote wurden neben der PTCA andere Therapieverfahren der invasiven Kardiologie entwickelt. Nachdem die intrakoronare Applikation diverser Pharmaka nicht zu einer nachweisbaren Prophylaxe der Restenosierung beigetragen hat (McBride et al. 1988; Blackshear et al. 1987; Whiteworth et al. 1986; Corcos et al. 1985), wird die Frage geprüft, ob mechanische Verfahren der Stenosenreduktion hier einen günstigeren Ansatzpunkt bieten. Im Erprobungsstadium befinden sich mehrere Verfahren zur Entfernung stenosierenden Materials aus dem Koronargefäß, so die Laserablation ohne (Karsch et al. 1989) und mit Wärmeanwendung („hot tip catheter") (Rosenthal et al. 1989), sowie unterschiedliche Atherektomieverfahren (Erbel et al. 1989; Fourrier et al. 1989; Simpson 1989). Mittels der „Stent"-Technik wird versucht, ein dauerhaftes mechanisches Hindernis für eine Restenosierung zu implantieren (Sigwart 1989; Schatz 1988; Sigwart et al. 1987). Sämtliche Verfahren sind von einer breiten klinischen Anwendung noch weit entfernt.

Unabhängig von den intrakoronar angewandten Revaskularisationstechniken sind durch die digitale Verarbeitung des primären Bildsignals und die Möglichkeit zur quantitativen Auswertung der digitalen Angiogramme erheblich verbesserte Grundlagen für die Befunderhebung geschaffen worden (Fleck et al. 1989). Die hier gezeigten Beispiele für Atherektomie und Perfusionskatheterangioplastien stellen Belege für die gegenwärtig erreichbare Qualität der Dokumentation dar.

Entwicklungsstufen der klassischen Angioplastie

Für die Anfangsphase der Angioplastie werden Perioden der technischen Entwicklung danach unterschieden, welche Instrumente eingesetzt werden konnten; so ist gekennzeichnet

Phase 1 durch die Verwendung von doppellumigen Ballonkathetern mit fixierten Drähten,
Phase 2 durch die Verwendung von steuerbaren Führungsdrähten,
Phase 3 durch die Verwendung von low-profile-Kathetern.

Diese Entwicklungen dienten vor allem dazu, den Primärerfolg der Therapie dadurch zu verbessern, daß die zu dilatierenden Stenosen sicherer sondiert und dilatiert werden konnten; gleichzeitig wurde die Inzidenz von Akutkomplikationen reduziert, wie die nachfolgende Übersicht zeigt (nach Tuzcu et al. 1989; Angaben in %):

Ergebnis	Phase 1 ($n=168$)	Phase 2 ($n=1117$)	Phase 3 ($n=1392$)
Stenose nicht sondiert	16,6	1,3	3,1
Stenose nicht dilatiert	3,0	0,2	0,4
Dissektion	4,8	1,5	1,4
Akuter Verschluß	3,3	2,5	1,4
Notfall-ACVB-Op.	8,3	4,2	2,5
Myokardinfarkt	7,1	3,3	2,4
Tod	0	0,2	0,4
Erfolgreiche Dilatation	72,6	94,5	93,8

Mit der Zunahme der technischen Möglichkeiten und der Erfahrungen werden zunehmend auch mehr Patienten mit erhöhtem Risiko für die Behandlung akzeptiert. Dies kann die Ergebnisstatistik späterer Zeiträume auch negativ beeinflussen (Meier u. Grüntzig 1984; Hartzler 1990). Sogar die Beherrschung von Komplikationen wurde durch diese technischen Entwicklungen erheblich verbessert. Während in der Frühphase praktisch jeder durch Angioplastie bewirkte Gefäßverschluß die dringliche operative Revaskularisation erforderlich machte, nahm diese Indikation seit Einführung der steuerbaren Führungsdrähte – und damit der Möglichkeit der unmittelbar wiederholten Angioplastie – merklich ab. Das PTCA-Register des NHLBI verzeichnete 1977/81 in 6,6 % der Fälle die Notfalloperation, 1985/86 jedoch nur noch in 3,9 % (Detre et al. 1988). Im eigenen Patientenkollektiv wurden zwischen 1986 und 1989 die folgende Inzidenz der Notfalloperation gefunden (Frantz et al. 1989):

Jahr	PTCA	Notfall-operation	[%]
1986[a]	156	4	2,56
1987	297	7	2,36
1988	445	7	1,57
1989	522	7	1,34
Gesamt	1420	25	1,76

[a] April–Dezember

Weiterentwicklung: Autoperfusionssysteme

Die zum Zweck der Dilatation bewirkte Insufflation des Ballonkatheters bewirkt einen vollständigen Verschluß des Koronargefäßes. Dieser Umstand limitiert die Zahl sowie Einzel- und Gesamtdauer der zur Dilatation aufzuwendenden Insufflationen; ggf. muß eher mit hohem Druck über kurze Zeit insuffliert werden. Voruntersuchungen haben demgegenüber ergeben, daß die Restenosierungsrate positiv beeinflußt wird, wenn über längere Zeit mit geringerem Druck insuffliert wird (Banka et al. 1989; Stack et al. 1988; Turi et al. 1987; Kaltenbach et al. 1984; Meier et al. 1984). In zahlreichen Fällen ist es erforderlich, einen beträchtlichen Anteil der Koronarperfusion im zu dilatierenden Gefäß zu erhalten; dies gilt insbesondere dann, wenn das zu dilatierende Gefäß das einzige für die Myokardperfusion zur Verfügung stehende bleibt sowie allgemein bei eingeschränkter linksventrikulärer Funktion. Schließlich ist die Dilatation einer Stenose des Hauptstammes der linken Kranzarterie nur bei erhaltener Restperfusion des Gefäßes denkbar (O'Keefe et al. 1989). Es war daher erforderlich, Dilatationssysteme zu entwickeln, bei denen eine solche Autoperfusion während der Balloninsufflation möglich ist.

Wir setzten bisher an 35 Patienten zur PTCA ein solches System ein (vgl. Abb. 2). Dilatiert wurde in 4 Fällen der Hauptstamm der linken Koronararterie, in 26 eine hauptstammnahe LAD-Stenose und in 5 Fällen eine anders lokalisierte Stenose. Die durchschnittliche Dauer der Insufflationen betrug 623 s (Spanne 360–1450 s), die verwendeten Ballongrößen betrugen ($\bar{x} \pm SD$) 3,44 ± 0,39 mm (Spanne 2,5–4,0 mm), die angewandten Insufflationsdrücke betrugen ($\bar{x} \pm SD$) 4,26 ± 0,74 atm[1] (Spanne 3,0–6,0 atm).

Die PTCA mittels Autoperfusionssystem konnte die folgenden Verbesserungen bewirken (quantitative Angiographie, angegeben sind Mittelwert ($\bar{x}$) ± Standardabweichung (SD)):

Parameter	Vor PTCA	Nach PTCA
Gefäßquerschnitt (mm^2)	0,82 ± 0,39	3,55 ± 1,88
Flächenreduktion (%)	92,4 ± 3,4	49,6 ± 10,6
Durchmesserreduktion (%)	73,1 ± 5,9	33,1 ± 10,7
Koronare Flußreserve	2,48 ± 0,76	4,87 ± 0,19

Bei 30 Patienten (85 %) wurde bereits eine Kontrollangiographie durchgeführt, über deren Ergebnisse hier berichtet wird (Krülls-Münch et al. 1989). Eine Restenosierung wurde dann bejaht, wenn am dilatierten Gefäßabschnitt wieder eine Stenose gefunden wurde, die mindestens 50 % des nichtstenosierten Gefäßdurchmessers betrug.

Bei 13 (43,3 % der mit Perfusionskatheter behandelten) Patienten wurde bei einem Follow-up von 2 Monaten eine so definierte Restenose nachgewiesen.

Die Parameter der dilatierten Gefäßabschnitte mit und ohne Restenose nach Dilatation durch Autoperfusionssysteme wurden verglichen (quantitative Angiographie). Es ergab sich ($\bar{x} \pm SD$):

[1] 1 atm = 101,3 kPa.

Parameter	Bei Gefäßen mit Restenose nach Dilatation	Bei Gefäßen ohne Restenose nach Dilatation
Durchmesser (mm)	1,39 ± 0,48	1,85 ± 0,45
Fläche (mm^2)	1,68 ± 1,24	2,82 ± 1,31
Durchmesserreduktion (%)	60,5 ± 22,9	20,7 ± 11,0
Flächenreduktion (%)	79,7 ± 14,7	35,2 ± 16,5
Koronare Flußreserve	3,28 ± 1,91	4,98 ± 0,06

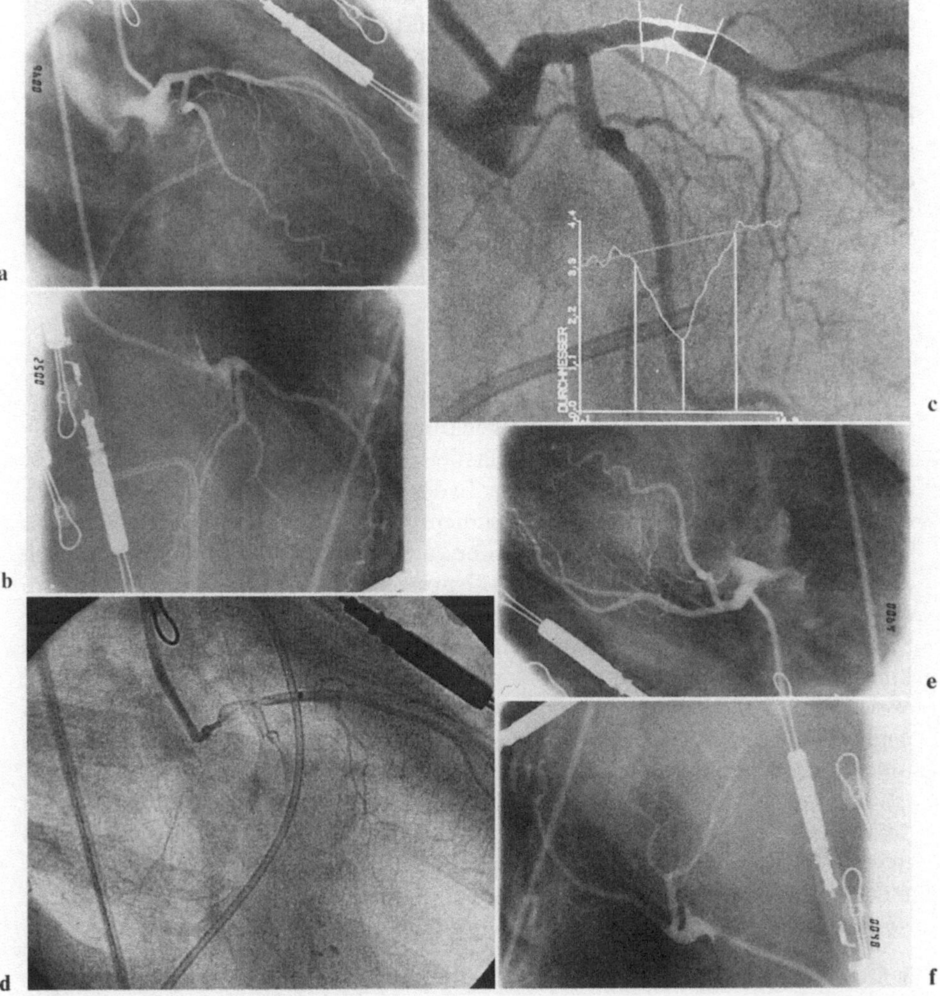

Abb. 2a–f. PTCA einer hochgradigen Stenose der LAD mittels Perfusionskatheter. **a** Angiographie vor PTCA in RAO-Projektion; **b** Angiographie vor PTCA in LAO-Projektion; **c** Quantitative Ermittlung der Durchmesserreduktion vor PTCA; **d** Angioplastie mittels Perfusionskatheter: Angiographie und Perfusion des Gefäßes bei insuffliertem Ballon; **e** Angiographie nach PTCA in RAO-Projektion; **f** Angiographie nach PTCA in LAO-Projektion

Die Daten zeigen, daß durch die Anwendung des Systems eine gegenüber der konventionellen Angioplastie erheblich verlängerte Perfusionszeit erreicht werden kann; die aufgewandten Drücke blieben erwarteterweise unterhalb von 6,0 atm. Suffiziente Primärergebnisse nach PTCA wurden erreicht.

In keinem Fall wurde durch die Dilatation mit dem Perfusionssystem ein Gefäßverschluß, ein Myokardinfarkt oder eine Komplikation ausgelöst, bei der eine Notfall-ACVB-Operation indiziert war. In 2 Fällen wurde durch den vorgegebenen Umfang des Führungskatheters eine zirkuläre Dissektion des Plaques am Abgang der LAD aus dem Hauptstamm bewirkt, die die elektive chirurgische Revaskularisation erforderte. Beide Patienten wurden 24 h nach der Behandlung mit gutem Ergebnis operiert.

Nach diesen Ergebnissen ist diese neue Technik also zur Dilatation von hauptstammnahen und im Hauptsstamm gelegenen Stenosen geeignet, ohne hierbei eine – gegenüber der PTCA eines „peripheren" Koronargefäßes – gesteigerte Akutkomplikationsrate aufzuweisen. Auch bei dieser Technik werden jedoch in beträchtlichem Maße Restenosierungen gefunden, ohne daß eine Identifikation der gefährdeten Gefäße oder Stenosen bisher gelungen wäre. Von weiteren Untersuchungen wird es abhängen, sicher zu beurteilen, welche Patienten von dieser Technik profitieren können.

Erweiterte Indikationsstellung für die PTCA

Die beschriebenen technischen Entwicklungen und erheblich erweiterte klinische Erfahrungen mit der Angioplastie haben es ermöglicht, Patienten für die Therapie zu akzeptieren, bei denen in der Frühphase die Gefährdung durch das Verfahren inakzeptabel hoch erschien. Während in der 1977/81-Kohorte des NHLBI-Registers lediglich 25 % der Patienten eine Mehrgefäßerkrankung aufwiesen, waren dies in der 1985/86-Gruppe schon 53 % der Patienten (Detre et al. 1988). Eine Auswertung von 494 Fällen von Mehrgefäßdilatation (1117 Stenosen) erbrachte einen Primärerfolg bei 89 % der Dilatationen, mit einer klinischen Verbesserung bei 95 % der Patienten (Myler et al. 1987). Diese Werte stimmen durchaus überein mit den entsprechenden Parametern bei Eingefäßdilatationen. Auch die Komplikationsrate und die Restenosierungsquote unterschieden sich nicht von anderen untersuchten Gruppen.

Die Auswertung des eigenen PTCA-Kollektivs 1986–1989 erbrachte bei 1420 Patienten in 25 Fällen die Indikation zur Notfall-ACVB-Operation (3 dieser Patienten waren aufgrund eines akuten Koronarverschlusses bereits vor PTCA-Beginn im kardiogenen Schock). Unter diesen Patienten waren die Stenosen und Dilatationsversuche wie folgt verteilt (Frantz et al. 1989):

1 Gefäß – 1 Stenose:	bei 18 Patienten,
1 Gefäß – 2 Stenosen:	bei 2 Patienten,
mehrere Gefäße:	bei 5 Patienten.

Auch diese Übersicht zeigt, daß die Mehrgefäß-/Mehrstenosendilatation nicht übermäßig häufig zu einem ungünstigen Primärergebnis führt.

In einer Übersicht über 6500 Dilatationen, bei der andere Hochrisikogruppen (Patienten in höherem Alter und mit eingeschränkter linksventrikulärer Funktion) ausgeschlossen waren, ergab sich kein Mortalitätsunterschied: von 1831 Patienten mit Eingefäßdilatation verstarben 5 (0,3 %), von 2385 Patienten mit Mehrgefäßdilatation ebenfalls 5 (0,2 %). Der Primärerfolg betrug in der Mehrgefäßgruppe 96 %, verglichen mit 88 % in der Eingefäßgruppe (Hartzler et al. 1988).

Diese Ergebnisse zeigen, daß es derzeit nicht gerechtfertigt ist, bei der Mehrgefäß-/Mehrstenosendilatation von einem gesteigerten Risiko des Mißerfolgs, der Akut- oder der Spätkomplikation auszugehen. Insofern ist die Indikation zur PTCA durch diesen Umstand nicht beschränkt. Unberührt bleibt, daß in Fällen der Mehrgefäßerkrankung häufiger eine eingeschränkte linksventrikuläre Pumpfunktion vorliegt, wodurch das Risiko des Verfahrens gesteigert wird.

Atherektomie in Koronargefäßen

Nachdem sowohl am Tiermodell als auch an stenosierten peripheren menschlichen Arterien (Simpson et al. 1989; Höfling et al. 1988) gezeigt werden konnte, daß mittels verschiedener Atherektomietechniken eine suffiziente und langfristig erfolgreiche Reduktion des stenosierenden Materials erzielt werden kann, wurden derartige Verfahren auch für die intrakoronare Anwendung am Patienten entwickelt.

Ein solches Verfahren scheint vor allem dann indiziert, wenn harte (fibrosierte oder kalzifizierte), exzentrische und durch Angioplastie nicht dilatierbare Stenosen behandelt werden sollen. Ziel dieses Verfahrens ist es, ein weiteres Lumen, eine glattere Oberfläche unter Vermeidung der bei der PTCA häufigen Wanddissektion zu erzeugen (Hinohara et al. 1989). Über die derzeit größte mit diesem Verfahren behandelte Gruppe berichtet Simpson, in dessen Gruppe der hier angewandte Atherektomiekatheter entwickelt wurde. Während unter den ersten 32 von ihm behandelten Patienten nur bei 55 % eine definitive Revaskularisation erzielt werden konnte, wurden von der zweiten 32-Patienten-Kohorte immerhin bereits 86 % definitiv behandelt. In dieser Kohorte wurde bei 28 Patienten Material entfernt, während dies nur bei 20 Patienten der ersten Kohorte gelungen war (Simpson 1989).

Simpson berichtet über Komplikationen bei 4 der oben erwähnten 64 Patienten: 1 Patient wurde (nach Dissektion und Gefäßverschluß) elektiv chirurgisch versorgt, zweimal war eine sofortige PTCA erforderlich und erfolgreich, eine Luftembolie löste sich spontan auf.

Dieses Verfahren wurde hier bisher an Nativkoronararterien von 8 Patienten mittels der von Simpson (1989) beschriebenen Technik angewandt (vgl. Abb. 3), 4 Patienten wurden bisher (3 Monate nach Durchführung der Prozedur) nachuntersucht. Die quantitative Angiographie erbrachte die nachstehend verzeichneten Ergebnisse:

Parameter des stenosierten Gefäßabschnitts	Vor TEA	Nach TEA	3 Monate nach TEA
Durchmesser (mm)	1,1 ± 0,5	2,5 ± 0,6	2,5 ± 1,1
– Reduktion (%)	68 ± 7,9	15 ± 9,4	–
Gefäßquerschnitt (mm^2)	1,0 ± 0,9	4,7 ± 2,3	5,7 ± 3,5
– Reduktion (%)	90 ± 5,5	27 ± 16	–
Koronare Flußreserve	2,7 ± 1,3	5,0 ± 0,1	–
Plaqueausdehnung (mm^2)	10,7 ± 8,8	–[a]	–[a]

[a] Nicht mehr quantitativ erfaßbar

In der eigenen Patientengruppe ließen sich also günstige Primärergebnisse erzielen; die Nachuntersuchungsergebnisse sind, soweit bereits vorliegend, ebenfalls positiv.

Eine (jüngere) Übersicht der Arbeitsgruppe Simpson et al. über 93 behandelte Stenosen bei 83 Patienten zeigt einen Primärerfolg (Reduktion der Stenose auf < 50 %) von 72 %. In dieser Patientengruppe war dreimal die elektive ACVB-Operation, einmal die PTCA indiziert, es ereigneten sich je ein transmuraler und ein nichttransmuraler Infarkt (Robertson et al. 1989 a).

Mittels Atherektomieverfahren wurde die Dilatation von Abgangsstenosen unternommen: Bei Behandlung von Ostien des Hauptstammes (n = 5) und von Venengrafts (n = 7) aus der Aorta kam es zu signifikant ungünstigeren Erfolgsquoten, größeren Reststenosen und häufigeren Dissektionen als bei Atherektomie am Ostium der LAD aus dem Hauptstamm (n = 14) (Robertson et al. 1989 b).

Die Primärerfolgsquote bei der Dilatation von Stenosen in Venengrafts beträgt am Ostium 71 %, bei proximalen oder distalen Schaftstenosen um 100 % (Selmon et al. 1989). Ein ungünstigeres Bild ergibt die Nachangiographie: Die Restenosierungsrate nach Atherektomie mit diesem Verfahren liegt nach (im Mittel) 5,3 Monaten bei Nativgefäßen um 27 %, bei Nativgefäßen mit zuvor durchgeführter mehrmaliger PTCA bei 47 %, bei Venengrafts – unabhängig von einer vorigen PTCA – um 55 % (Simpson et al. 1989).

Ein zusätzlicher Nutzen dieses Verfahrens liegt darin, entferntes stenosierendes Material für weitere Untersuchungen zu gewinnen (v. Pölnitz et al. 1990; Liu et al. 1989; Lancet 1989; Johnson et al. 1988). Hierbei wird Aufschluß über die unterschiedliche pathologische Anatomie von primär atheromatösen Plaques einerseits und klinisch als „Restenosierung" auftretendem Material andererseits erwartet.

Dieses Material ist damit nicht nur der histologischen Untersuchung zugänglich; vielmehr besteht die Möglichkeit der Anlage von Zellkulturen, aus denen wiederum Einflußfaktoren für Proliferation und Wachstumshemmung abgeleitet werden können (Dartsch et al. 1989). Damit könnten ggf. klinisch relevante Behandlungsmöglichkeiten zur Prävention von Restenosierung nach PTCA abgeleitet werden.

Zusammenfassend zeichnet sich diese Atherektomiemethode also durch günstige Primärergebnisse aus, während die Restenosierungsrate in den ersten

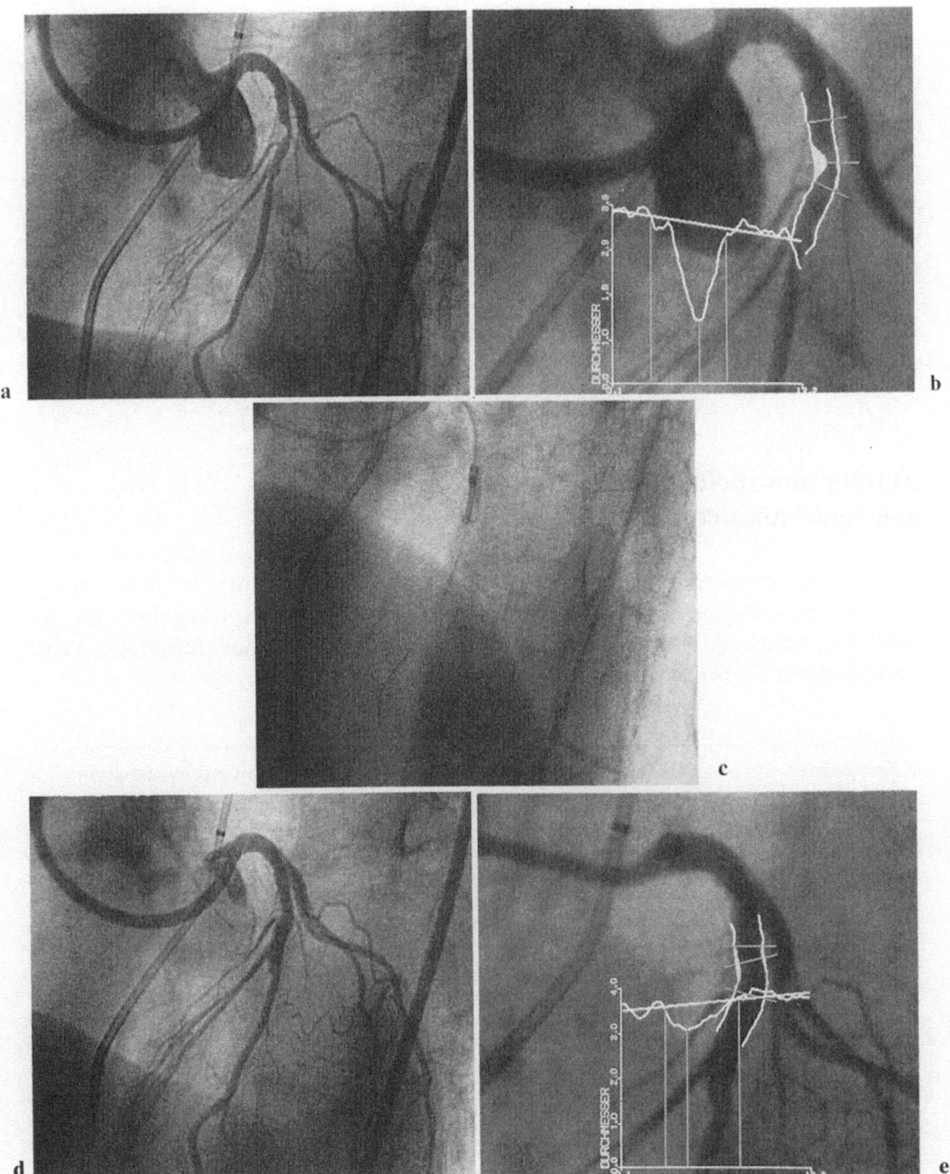

Abb. 3a–e. Atherektomie einer hochgradigen Stenose der LAD mittels Simpson-Katheter. **a** Angiographie vor Atherektomie in LAO-Projektion; **b** Quantitative Ermittlung der Durchmesserreduktion vor Atherektomie; **c** Instrument in situ; **d** Angiographie nach Atherektomie in LAO-Projektion; **e** Quantitative Ermittlung der Durchmesserreduktion nach Atherektomie

behandelten Patientengruppen bisher noch schlechter als bei der konventionellen PTCA erscheint. Die Restenosierungsrate scheint mit der Zahl der vorherigen Manipulationen an der Stenose zuzunehmen. Ostiumnahe und Stenosen in Venengrafts können mittels dieser Methode mit gutem Primärerfolg angegangen werden, wobei die Atherektomie an Ostien in der Aorta den geringsten Primärerfolg aufweist. Die Methode hat zu Komplikationen geführt, die eine elektive chirurgische Intervention und die dringliche Angioplastie erforderlich machten. Vor allem die Embolisation durch abgetragenes stenosierendes Material stellt eine neuartige Komplikation dar.

Über den Revaskularisationszweck hinaus beinhaltet die Methode die Gelegenheit, Material aus nativen Stenosen und Restenosierungen für das Studium der zugrundeliegenden Mechanismen zu gewinnen.

Koronarmorphologie als Prädiktor für Akut- und Spätkomplikationen

Die Literaturübersicht (Frantz et al. 1990) über neuere Studien, in denen die Koronarmorphologie in bezug auf ein ungünstiges Therapieergebnis ausgewertet worden ist, ergibt für die folgenden Kriterien einen möglicherweise ungünstigen Einfluß:

Kriterium	Patienten in Studien mit nachweisbarem Einfluß auf das Risiko zur			
	Akutkomplikation		Restenosierung	
	n	[%]	*n*	[%]
Druckgradient nach PTCA	4712	100	1706	71,2
Exzentrizität der Stenose	8127	97,1	345	17,2
Gefäßverzweigung	5072	95,4	–	
Gefäßbiegung	4472	94,9	–	
Thrombus	4472	94,9	–	
% Stenose nach PTCA	4782	93,5	1687	55,8
Dissektion	4782	90,1	–	
Länge der Stenose	4472	85,4	496	29,6
% Stenose vor PTCA	3079	40,4	1053	45,1
Druckgradient vor PTCA	–		557	75,5
Stenose in LAD	–		1769	52,7
Kalzifizierung	–		181	10,6

In mehreren Studien hatten die Kriterien „weitere Stenose im Koronargefäß", „Ulzeration", „abrupter Stenosenbeginn", „unregelmäßiges Lumen" keinen nachweisbaren Einfluß auf beide Risiken. Die unterschiedliche Verteilung in beiden Komplikationsgruppen zeigt: Für die Akutkomplikationen besteht ein klares Bild hinsichtlich der zum Problem führenden Koronaranatomie, für die Restenosierung ist eine solche Identifikation derzeit nur näherungsweise möglich.

PTCA bei Hochrisikopatienten

Nachdem Instrumente zur Verfügung stehen, die eine Perfusion des dilatierten Gefäßes während der Dilatation erlauben und die Mehrstenosen-PTCA per se nicht mit hohem Risiko behaftet scheint, ergibt sich die Frage, in welchen Fällen nach heutigem Kenntnisstand von einem erhöhten Risiko auszugehen ist.

Als Risikogruppen werden (nach Hartzler et al. 1988) Patienten mit akutem Myokardinfarkt, eingeschränkter linksventrikulärer Funktion, Hauptstammstenose (oder äquivalenter Koronaranatomie), höherem Alter und vorangegangener Bypass-Chirurgie angesehen.

Im akuten Myokardinfarkt ist durch eine Vielzahl von Studien Indikation, Erfolg und Risiko der systemischen und intrakoronaren Lysetherapie mit verschiedenen Thrombolytika untersucht worden (Schröder 1989). Der positive Einfluß einer solchen Therapie steht – soweit Kontraindikationen beachtet und ausgeschlossen werden – außer Zweifel. Eine intensive Diskussion ist über die Frage geführt worden, ob eine PTCA nach erfolgreicher Lysetherapie in möglichst kurzem zeitlichem Abstand angestrebt werden sollte.

Einhelliges Ergebnis dreier großer Studien (TIMI II Study Group 1989; Simoons et al. 1988; TAMI Study Group 1987) ist es, daß zwar eine PTCA den auf pharmakologischem Weg erreichten Revaskularisationserfolg sichern kann, daß aber mit einer Notfall-PTCA gegenüber der späteren elektiven PTCA einerseits kein Vorteil hinsichtlich des Primärerfolgs bewirkt wird und andererseits eine erhebliche Steigerung der Angioplastie- und Punktionsrisiken verursacht wird.

In einem Bericht über 446 PTCA bei Patienten im akuten Myokardinfarkt konnte eine Primärerfolgsquote von 93 % verzeichnet werden, bei einer Hospitalmortalität von 8,5 %. Bei Patienten mit nur gering eingeschränkter linksventrikulärer Pumpfunktion betrug die Mortalität 1,5 %, bei solchen mit LV-EF < 30 % allerdings 38 % (Hartzler et al. 1988). Die PTCA hat beim akuten Myokardinfarkt also insbesondere in den Fällen einer gegen die Lysetherapie bestehenden Kontraindikation ihre Indikation (Kander et al. 1989); nach erfolgreicher Lyse sollte die PTCA jedoch eher elektiv und nicht unmittelbar durchgeführt werden. Kontrollierte Studien zum direkten Vergleich von Erfolg und Risiken pharmakologischer, mechanischer (PTCA) und chirurgischer Revaskularisation im akuten Myokardinfarkt liegen derzeit noch nicht vor (Faxon 1988; King 1988).

Nachfolgende Daten geben einen zusammenfassenden Überblick über Erfolg und ungünstigen Ausgang der PTCA bei Hochrisikogruppen (aus 6500 PTCA 1980–1987, nach Hartzler et al. 1988):

Risikogruppe	n	Primär-Erfolg [%]	Myokard-Infarkt [%]	Notfall-ACVB-Op. [%]	Tod [%]	Risiko-faktor
Hauptstammstenose	103	92	0,9	1,9	3,9	5,9
Hauptstammstenose-Äquivalent	77	95	1,3	6,5	2,6	3,8
LV-EF <40%	664	93	0,7	2,0	2,7	5,9
Alter >70 Jahre	1038	94	0,8	1,3	1,4	2,6
Akuter Myokardinfarkt	446	93		2,9	8,5	11,2
Akuter Myokardinfarkt + LV-EF <40%	63				38	52,9

Konsequenzen

Die Beobachtung der umfangreichen Daten zu Erfolgs-, Komplikations- und Restenosierungsrate nach PTCA hat neben der Verbesserung durch die dargestellten technischen Entwicklungen auch den Einfluß der wachsenden Erfahrung der Operateure mit der Methode deutlich gemacht. Aus dieser generellen Betrachtung ist das Bestehen einer sog. „learning curve" abgeleitet worden (Meier u. Grüntzig 1984), die angibt, daß nach Ende der Frühzeit dieser Methode seit mehreren Jahren jetzt die Erfolgsquote und Komplikationshäufigkeit bei PTCA sich konstant verhalten. Hieraus ist zu schließen, daß die Beurteilung der Erfolgsquote eines Zentrums oder eines Operateurs sich nicht an früheren Ergebnissen, sondern am derzeit üblichen Standard zu orientieren hat.

Damit ein solches konstantes Ergebnis erzielt wird, ist es erforderlich, daß Operateure die Behandlung nur ihren Erfahrungen entsprechend übernehmen. Vorausgesetzt wird, daß eine hinreichende praktische und theoretische Ausbildung bezüglich Koronarangiographie (mindestens 1000 selbständig durchgeführte Untersuchungen) und -angioplastie unter Aufsicht erfahrener Operateure stattgefunden hat. Auf dieser Basis ist es möglich anzugeben, welche Indikationen dem Erfahrungsstand des Operateurs entsprechen (nach Hartzler 1990):

Operateur	Anzahl der PTCA	Indikationen
– Anfänger	<100	Eingefäßerkrankung/Einstenosendilatation Linksventrikuläre Ejektionsfraktion <45% Stenosenlänge <20 mm
– fortgeschritten	100–500	Mehrgefäßerkrankung Mehrstenosendilatation Wiedereröffnung chronisch okkludierter Gefäße Intervention im akuten Myokardinfarkt Linksventrikuläre Ejektionsfraktion <45%

Operateur	Anzahl der PTCA	Indikationen
– erfahren	>500	Angioplastie des Hauptstamms (oder -Äquivalent) Hohes Lebensalter (±eingeschränkte LV-Funktion) Intervention im kardiogenen Schock Inoperable Patienten

Zusammenfassung

Durch technische Entwicklung und klinisch erweiterte Erfahrungen hat sich die PTCA zum weitestverbreiteten Instrument der nichtchirurgischen Revaskularisation bei koronarer Herzerkrankung entwickelt. Zahlreiche weitere Techniken der mechanischen Manipulation im stenosierten Koronargefäß werden entwickelt, eine der PTCA vergleichbare klinische Verbreitung haben diese Verfahren bisher noch nicht gefunden.

Insbesondere durch Entwicklung von Autoperfusionssystemen ist nunmehr auch die Dilatation von hauptstammnahen und im Hauptstamm gelegenen Stenosen möglich geworden. Dies und die breite Erfahrung mit der Dilatation von mehreren Stenosen bei Mehrgefäßerkrankung sowie mit der Wiedereröffnung chronisch okkludierter Gefäße hat die Indikation zur PTCA erheblich erweitert.

Die Akutkomplikation durch plötzlichen Gefäßverschluß sind nach derzeitiger Übersicht mit größerer Wahrscheinlichkeit vorherzusagen als die später auftretende Restenosierung nach PTCA. Solche Komplikationen erfordern nach wie vor die sofortige chirurgische Intervention zur definitiven Revaskularisation mittels aortokoronarer Venengrafts.

Pathophysiologische Mechanismen der Restenosierung werden bereits eingehend untersucht – die Atherektomie aus Koronararterien hat darüber hinaus die Möglichkeit ergeben, entsprechendes Zellmaterial in vitro zu untersuchen bzw. anzuzüchten. Eine klinisch verwertbare Identifikation der Patienten bzw. der Gefäße und Stenosen, die diesem Risiko besonders häufig unterliegen, hat sich derzeit noch nicht ergeben.

Literatur

Banka VS, Trivedi A, Patel R, Ghusson M, Voci G (1989) Prevention of myocardial ischemia during coronary angioplasty: A simple new method for distal antegrade arterial blood perfusion. Am Heart J 118:830–836

Blackshear JL, O'Callaghan WG, Califf RM (1987) Medical approaches to prevention of restenosis after coronary angioplasty. J Am Coll Cardiol 9:834–848

Corcos T, David PR, Val PG, Renkin J, Dangoisse V, Rapold HG, Bourassa MG (1985) Failure of diltiazem to prevent restenosis after percutaneous transluminal coronary angioplasty. Am Heart J 109:926–931

Dartsch P, Bauriedel G, Höfling B, Betz E (1989) Cell culture of human atheromatous plaque. In: Höfling B, Pölnitz A von (eds) Interventional cardiology and angiology. Steinkopff, Darmstadt

Detre K, Holubkov R, Kelsey S, Cowley M, Kent K, Williams D, Myler R, Faxon D, Holmes D, Bourassa M, Block P, Gosselin A, Bentivoglio L, Leatherman L, Dorros G, King SB III, Galicha J, Al-Bassam M, Leon M, Robertson T, Passamani E (1988) Percutaneous transluminal coronary angioplasty in 1985–1986 and 1977–1981. The National Heart, Lung and Blood Institute Registry. New Engl J Med 318:265–270

Ellis SG, Shaw RE, Gershony G, Thomas R, Roubin GS, Douglas JS, Topol EJ, Startzer SH, Myler RK, King SB III (1989) Risk factors, time course and treatment effect for restenosis after successful percutaneous transluminal coronary angioplasty of chronic total occlusion. Am J Cardiol 63:897–901

Erbel R, O'Neill W, Auth D, Haude M, Nixdorff U, Dietz U, Rupprecht HJ, Tschollar W, Meyer J (1989) Hochfrequenz-Rotationsatherektomie bei koronarer Herzkrankheit. Dtsch Med Wochenschr 114:487

Faxon DP (1988) The risk of reperfusion strategies in the treatment of patients with acute myocardial infarction. J Am Coll Cardiol 12 [Suppl. A]: 52A–57A

Fleck E, Dirschinger J, Rudolph W (1985) Quantitative Koronarangiographie vor und nach PTCA. Herz 10:313–320

Fleck E, Regitz V, Lehnert A, Dacian S, Dirschinger J, Rudolph W (1988) Restenosis after balloon dilatation of coronary stenosis – multivariate analysis of potential risk factors. Eur Heart J 1988, 9 [Suppl. C]:15–18

Fourrier JL, Bertrand ME, Auth DC, Lablanche JM, Gommeaux A, Brunetaud JM (1989) Percutaneous coronary rotational angioplasty in humans: preliminary report. J Am Coll Cardiol 14:1278–1282

Frantz E, Krülls-Münch J, Oswald H, Strasser R, Fleck E (1989) Notfall-Operation nach PTCA: Koronar-Anatomie, geplante und durchgeführte Therapie (Abs.). Z Kardiol 78 [Suppl. 4]:22

Frantz E, Krülls-Münch J, Oswald H, Fleck E (1990) Emergency CABG – surgery after PTCA, complications in 1420 cases of angioplasty. In: Fleck E, Frantz E (eds) Complications in PTCA, Steinkopff, Darmstadt (in press)

Gleichmann U, Mannebach H (1989) 5. Bericht über die Leistungszahlen der Herzkatheterlabors in der Bundesrepublik Deutschland, Z Kardiol 78:811–817

Hartzler GO (1990) Final remarks: prevention of complications by angioplasters' training and patient-selection, in: Fleck E, Frantz E (eds) Complications in PTCA. Steinkopff, Darmstadt (in press)

Hartzler GO, Rutherford BD, McConahay DR, Johnson WL (1988) "High risk" percutaneous transluminal coronary angioplasty. Am J Cardiol 61 [Suppl. G]:33G–37G

Hinohara T, Selmon M, Robertson G, White N, Rowe M, Simpson J (1989) Angiographic appearances following percutaneous coronary atherectomy (Abs.). J Am Coll Cardiol 13 [Suppl. A]:223A

Höfling B, Pölnitz A von, Backa D, Arnim TH von, Lauterjung L, Jauch KW, Simpson JB (1988) Percutaneous removal of atheromatous plaques in peripheral arteries. Lancet I:384–386

Holmes DR, Vliestra RE, Smith HC, Vetrovec GW, Kent KM, Cowley MJ, Faxon DP, Grüntzig AR, Kelsey SF, Detre KM, VanRaden MJ, Mock MB (1984) Restenosis after percutaneous transluminal coronary angioplasty (PTCA): A report from the PTCA registry of the National Heart, Lung, and Blood Institute. Am J Cardiol 53 [Suppl. C]:77C

Johnson DE, Robertson G, Simpson JB (1988) Coronary atherectomy: eight microscopic and immunohistochemical study of excised tissues (abs.) Circulation 78:II–82

Kaltenbach M, Beyer J, Walter S, Klepzig H, Schmidts L (1984) Prolonged application of pressure in transluminal coronary angioplasty. Cathet Cardiovasc Diagn 10:213–219

Kander NH, O'Neill W, Topol E, Gallison L, Mileski R, Ellis SG (1989) Long-term follow-up of patients treated with coronary angioplasty for acute myocardial infarction. Am Heart J 118:228–233

Karsch KR, Haase KK, Mauser M, Ickrath O, Voelker W, Duda S (1989) Percutaneous coronary excimer laser angioplasty: initial clinical results, Lancet II: 647–650

King SB III (1988) Percutaneous transluminal coronary angioplasty: the second decade. Am J Cardiol 62: 2K–6K

Krülls-Münch J, Oswald H, Alfakhri N, Fleck E (1989) PTCA von Hauptstammstenosen und hauptstammnahen Stenosen (Abs.). Z Kardiol 78 [Suppl. 4]: 14

Lancet (Editorial) (1989) Pathology of coronary angioplasty. Lancet II: 423–424

Lange RA, Hillis LD (1989) Short- and long-term complications of coronary angioplasty. Chest 96: 151–157

Liu MW, Roubin GS, King SB III (1989) Restenosis after coronary angioplasty. Circulation 79: 1374–1387

McBride W, Lange RA, Hillis LD (1988) Restenosis after successful coronary angioplasty. New Engl J Med 318: 1734–1737

McDonald RG, Panusch RS, Pepine CJ (1987) Rationale for use of glucocorticoids in modification of restenosis after percutaneous transluminal coronary angioplasty. Am J Cardiol 60 [Suppl. B]: 56B

Meier B, Grüntzig AR (1984) Learning curve for percutaneous transluminal coronary angioplasty: skill, technology or patient selection. Am J Cardiol 53 [Suppl. C]: 65C–66C

Meier B, Grüntzig AR, King SB III, Douglas JS, Hollmann J, Ischinger T, Galan K (1984) Higher balloon dilatation pressure in coronary angioplasty. Am Heart J 107: 619–622

Myler RK, Topol EJ, Shaw RE, Stertzer SH, Clark DA, Fishman-Rosen J, Murphy MC (1987) Multiple vessel coronary angioplasty: classification, results and patterns of restenosis in 494 consecutive patients. Cathet Cardiovasc Diagn 13: 1–15

O'Keefe JH, Hartzler GO, Rutherford BD, McConahay DR, Johnson WL, Giorgi IV, Ligon RW (1989) Left main coronary angioplasty: early and late results of 127 acute and elective procedures. Am J Cardiol 64: 144–147

Pölnitz A von, Bauriedel G, Dartsch PC, Schinko I, Backa D, Betz E, Welsch U, Höfling B (1990) Diagnostic implications of percutaneous atherectomy: angioscopic, histologic and cell culture study. In: Fleck E, Frantz E (eds) Complications in PTCA. Steinkopff, Darmstadt (in press)

Quigley PJ, Hinohara T, Phillips HR, Peter RH, Behar VS, Kong Y, Sionton CA, Perez JA, Stack RS (1988) Myocardial protection during coronary angioplasty with an autoperfusion balloon catheter in humans. Circulation 78: 1128–1134

Robertson G, Hinohara T, Selmon M, Rowe M, White N, Simpson J (1989a) Complications: early experiences of percutaneous coronary atherectomy (abs.). J Am Coll Cardiol 13 [Suppl. A]: 222A

Robertson G, Hinohara T, Selmon M, White N, Simpson J (1989b) Effectiveness of directional coronary atherectomy for the treatment of ostial lesions (abs.). Circulation 80 [Suppl. II]: II–582

Rosenthal E, Montarello JK, Palmer T, Curry PVL (1989) Coronary artery thermal damage during percutaneous "hot tip" laser-assisted angioplasty. Am J Cardiol 64: 116–120

Schatz RA (1989) A view of vascular stents. Circulation 79: 445–457

Schröder R (1989) Thrombolyse beim akuten Myocardinfarkt: Eine Standortbestimmung 1988. Z Kardiol 78: 41–62

Selmon MR, White NW, Sipperly ME, Robertson GC (1989) Efficacy and safety of directional coronary atherectomy for the treatment of bypass graft stenoses (abs.). Circulation 80 [Suppl. II]: II–583

Sigwart U (1989) Current problems with coronary stenting. Eur Heart J 10: 408

Sigwart U, Puel J, Mirkovitch V, Joffre F, Kappenberger L (1987) Intravascular stents to prevent occlusion and restenosis after transluminal angioplasty. New Engl J Med 316: 701–706

Simoons ML (1989) Reocclusion/restenosis after coronary artery bypass surgery, percutaneous transluminal coronary angioplasty and thrombolysis. Z Kardiol 78 [Suppl. 3]: 35–41

Simoons ML, Arnold AER, Betriu A, deBono DP, Dougherty FC, von Essen R, Lambertz H, Lubsen J, Meier B, Raynaud P, Rutsch W, Sanz GA, Schmidt W, Serruys PW, Thery C, Uebis R, Vahanian A, van de Werf F, Willems GM, Wood D, Verstraete M (1988) Thrombolysis with tissue plasminogen activator in acute myocardial infarction: no additional benefit from immediate percutaneous coronary angioplasty. Lancet I:197–203

Simpson JB (1989) Percutaneous transluminal atherectomy. In: Höfling B, Pölnitz A von (eds) Interventional cardiology and angiology. Steinkopff, Darmstadt

Simpson JB, Robertson GC, Selmon MR (1988) Percutaneous coronary atherectomy (abs.). Circulation 78 [Suppl. II]: II–82

Simpson JB, Robertson GC, Selmon MR, Sipperly ME, Braden LJ, Hinohara T (1989) Restenosis following successful directional coronary atherectomy (abs.). Circulation 80 [Suppl. II]: II–582

Stack RS, Quigley PJ, Collins G, Phillips HR (1988) Perfusion balloon catheter. Am J Cardiol 61:77–80

TAMI Study Group (1987) A randomized trial of immediate versus delayed elective angioplasty after intravenous tissue plasminogen activator in acute myocardial infarction. New Engl J Med 317:581–588

TIMI II Study Group (1989) Results of the thrombolysis in myocardial infarction (TIMI) phase II trial. New Engl J Med 320:618–627

Turi ZG, Campbell CA, Gottimukkala MV, Kloner RA (1987) Preservation of distal coronary perfusion during prolonged balloon inflation with an autoperfusion angioplasty catheter, Circulation 75:1273–1280

Tuzcu EM, Simpfendorfer C, Dorosti K, Franco I, Hollman J, Badhwar K, Whitlow P (1989) Changing patterns in percutaneous transluminal coronary angioplasty. Am J Heart J 117:1374–1377

Whitworth HB, Roubin GS, Hollman J, Meier B, Leimgruber PP, Douglas JS, King SB III, Grüntzig AR (1986) Effect of nifedipine on recurrent stenosis after percutaneous transluminal coronary angioplasty. J Am Coll Cardiol 8:1271–1276

Recent Advances in Excimer Laser Angioplasty

W. S. Grundfest, C. Beeder, J. Segalowitz, and F. Litvack

Introduction

Excimer laser angioplasty provides a means to open totally obstructed blood vessels, treat diffuse coronary artery disease and work in ever smaller arteries. This article will present a discussion of some of the techniques for, as well as the results of percutaneous laser interventions.

Developmental Background and Technical Aspects

Percutaneous transluminal coronary angioplasty (PTCA) has revolutionized the treatment of vascular disease, but unfortunately is not applicable to all patients. Laser angioplasty has the potential to extend the boundaries of nonsurgical revascularization. As such, it holds tremendous promise; however, this promise was not immediately apparent when we began our research in the early 1980s. Indeed, we encountered significant limitations in our initial trials with continuous wave lasers. We could perform recanalization, but only in noncalcified tissue. Perforation, reocclusion and subintimal dissection complicated all of the early studies. It was clear that success depended on a better understanding of how laser energy interacts with tissue, how the energy should be delivered and what type of delivery system to use.

We began using an excimer laser for several reasons: it has a short pulse duration, high energy per pulse and, at 308 nm, very intense tissue absorption. Indeed, the excimer laser can be focused so tightly and precisely that nuclei can be cut in half without damaging the corresponding tissue. Also, the excimer laser cuts calcified tissue. Consequently, we felt that the excimer laser had unique advantages in intravascular applications. It ablates atheroma with microscopic precision and great control. There is no thermal injury to the surrounding tissue and, more importantly, our studies showed that vessels healed with minimal scarring.

Unfortunately, when we first began using the excimer laser it was not a particularly practical device, if for no other reason than its size and immobility. However, we developed a strong collaborative effort with Dr. Laudenslager to produce another laser designed for clinical use. This new machine can be rolled between our research laboratory and the hospital's catheterization laboratories. It has a self-contained chiller so it does not need water cooling and it plugs directly into standard wall sockets. More importantly, through a great deal of work at Advanced Interventional Systems, with Dr. Tsui Goldenberg we developed the first fiber systems. The system is 4.5 m long and can transmit more than enough energy to cut granite. Of course, we never need that much energy in clinical situations, in which 35–60 mJ/mm is delivered to the tissue.

Subsequently we have developed systems inside catheters that use a single fiber with a centering balloon, as well as a system with twelve 200 µm fibers surrounding a central guide wire. Alignment of the fiber system at its connection to the laser is simple and accomplished by simply screwing the fiber into a coupler.

The single fiber that we first used in our patients was very crude and was really designed to make a hole in a total obstruction. It was quite rigid, which, for the peripheral arteries, was tolerable. The fiber transmits more than enough energy to make a 1.2 mm hole. In contrast, the over-the-wire system with the twelve 200 µm fibers tracks the fiber, a is very flexible and creates a 1.6 mm channel. It is also possible to perform distal dye injections with the wire in place, so we can use concurrent angiography. We can also put a small angioscope down the lumen if desired.

Clinical Results after Excimer Laser Therapy

We have developed a clinical data base in which all patients are classified by symptoms, angiography, ankle-brachial index and color flow Doppler. We use the color flow Doppler to precisely locate the channel of the artery and the origin of the occlusion. It also shows the presence of collaterals and whether there is anterograde or retrograde flow. All patients have a detailed preoperative hospital evaluation and receive aspirin. Follow-up occurs at 1, 2, 4, 6 and 12 months. To date, no patient has been lost to follow-up.

In our first 30 patients we have treated 9 stenoses and 22 occlusions. When inserting the fiber we are continually injecting serial puffs of dye to locate the position of the fiber in relation to the artery. We now also use a road mapping technique with digital angiography and we find that very helpful, as we can detect the presence of the fiber as it goes down. The other field that we are looking into is the use of Doppler to help give us a more three-dimensional picture. The key to most problems now is guidance of the fiber. We have solved the problem of cutting: we must now solve the problem of guidance.

In peripheral arteriosclerosis, excimer laser therapy has been successful, with a recanalization rate of about 85% in stenoses of up to 15 cm in length. Our overall success rate is about 77% (Table 1). Interestingly, in our failures the primary reason for failure was inability to maintain a coaxial position. We think we have a partial solution to this problem now with a coaxializing balloon.

We have had some complications, most of them minor in nature (Table 2). The renal failure resolved without any sequelae, the embolus was removed by suction embolectomy and we have only one patient requiring bypass surgery.

Table 1. Excimer angioplasty of lower extremities

Acute success	n (%)
Stenoses	7/9 (78)
Occlusions	
0– 5 cm	6/7 (86)
6–10 cm	7/8 (88)
11–15 cm	3/4 (75)
>15 cm	1/3 (33)
Total	17/22 (77)
Overall success	24/31 (77)

Table 2. Excimer angioplasty of lower extremities

Complications	n
Groin hematoma	4
Tip detachment	1
Renal failure	1
Embolus	1
Bypass surgery	1
Early reocclusion	1

Of the 20 patients who were successfully dilated we have had four restenoses after a mean follow-up period of 9 months. All four of these patients were treated successfully by PTCA: so what we have done is to convert long total occlusions into short occlusions that can be dilated with a balloon. Today, the 20 patients whose arteries we opened all have patent vessels.

To date, 81 patients have undergone percutaneous excimer laser coronary angioplasty in a multicenter trial. Acute recanalization was obtained in 84% of these patients, with no deaths or perforations, although one patient urgently required coronary artery bypass graft (CABG, Margolis et al. 1989). Mean percent stenoses were reduced significantly, from a baseline of 89%, falling to 49% post laser and 22% after PTCA.

Reference

Margolis JR, Litvack F, Grundfest W et al. (1989) Excimer laser angioplasty: results of a multicenter study. Circulation 80 [Suppl II]: 477

Laserangioplastie und andere in der Entwicklung befindliche Rekanalisationstechniken

G. Biamino, H. Kar, K. Dörschel, P. Skarabis, M. Gross, G. Stefan, H. Böttcher, U. Flesch, H. Witt und G. Müller

Einleitung

Die perkutane transluminale Angioplastie hat im Laufe der letzten 10 Jahre in Verbindung mit der verbesserten Kathetertechnik eine weltweite Verbreitung erfahren. Die Akzeptanz dieser desobliterierenden Technik widerspiegelt sich in der Tatsache, daß 1989 weltweit etwa 320000 Interventionen im Koronarbereich vorgenommen wurden, davon etwa 15000 in der Bundesrepublik Deutschland. Die Anzahl der peripheren Dilatationen läßt sich schlecht abschätzen, dürfte jedoch ebenfalls im Bereich von 300000 Interventionen liegen.

Trotz dieses beachtlichen Siegeszuges bemühen sich weltweit etwa 30 Firmen um die Entwicklung von neuen Technologien zur Beseitigung von atherosklerotischen Ablagerungen in peripheren und koronaren Gefäßen. Die erforderlichen Interventionen wären sicherlich nicht sinnvoll, wenn die neuen Technologien nur dann eingesetzt werden könnten, wenn die Ballondilatation primär versagt: lediglich in 10–15% im Falle einer Stenose.

Der entscheidende Nachteil der koronaren und peripheren Ballondilatation besteht darin, daß mit einer relativ hohen Restenosierung bzw. Reokklusionsrate zu rechnen ist; bei peripheren Gefäßen liegt sie, insbesondere, wenn es sich um langstreckige Verschlüsse handelt, im Bereich bis zu 50–60%, im Koronarbereich bei 30% [4, 19, 20].

Sollte es mit Hilfe von neuen Technologien gelingen, die Restenosierungsrate zu halbieren, so könnte sich die entsprechende Methode wahrscheinlich neben der herkömmlichen PTA-Technik etablieren.

Es wird weiterhin von den neuen Techniken erwartet, daß die Primärerfolgsrate bei kompletten chronischen Verschlüssen von bisher knapp 50% erheblich gesteigert werden kann.

Futuristischen Vorstellungen amerikanischen Studien zufolge (Abb. 1) könnte man dann davon ausgehen, daß die Laserangioplastie und die mechanischen Rekanalisationsverfahren eine wesentliche Erweiterung der Indikation zur aktiven Intervention bei sklerotischen Obliterationen darstellen würden. Hierdurch könnte nicht nur partiell die konservative medikamentöse Therapie ersetzt werden, sondern auch deren Kombination mit chirurgischen Maßnahmen eine wesentliche Erweiterung des Indikationsfeldes erfahren.

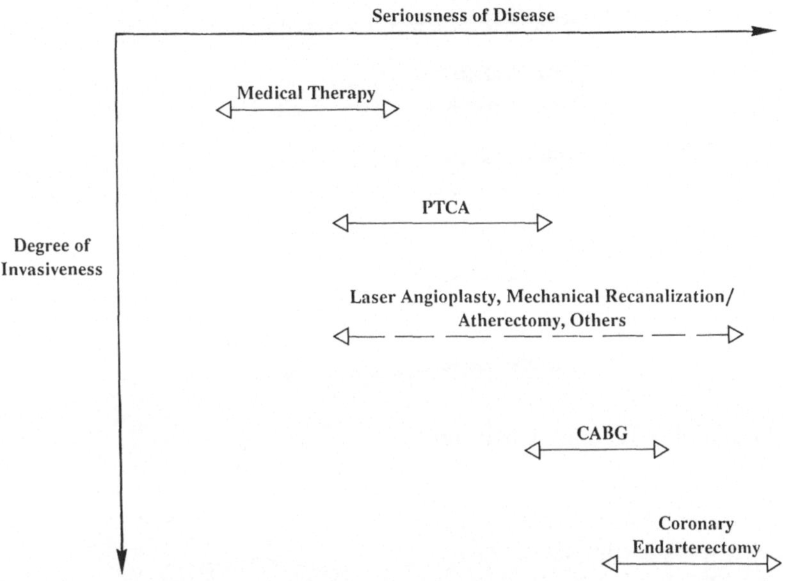

Abb. 1. Futuristische Vorstellungen über den Stellenwert der mechanischen Rekanalisationsverfahren und der Laserangioplastie (CV/MS Profile, Mai 1988)

Mechanische Rekanalisationstechniken

Eine Reihe von mechanischen Systemen befinden sich zur Zeit in der klinischen Erprobung bzw. sind bereits auf dem Markt erhältlich.

Das System, das sich bisher am meisten durchgesetzt hat, ist der Simpson-Directional-Atherectomy-Katheter (Abb. 22a) [25].

Weltweit dürften mit dieser Methode im peripheren Bereich bereits 700 Patienten behandelt worden sein, während im koronaren Bereich die Ergebnisse von etwa 300 Behandlungen vorliegen dürften (Höfling/München, persönl. Mitteilung).

Der AtheroCath-Katheter ist mit einem festen Führungsdraht von 4,5 cm versehen. Unmittelbar hinter der Katheterspitze befindet sich ein Metallzylinder mit einem offenen Fenster. Voraussetzung für die Anwendung dieses Systems ist, daß mit dem Katheter die Stenose passiert werden muß. Zur Entfernung des atherosklerotischen Materials wird der Katheter so plaziert, daß das Fenster im Metallgehäuse das abzutragende Material soweit wie möglich einschließt. Diese „direktionale Positionierung" des Katheters wird durch einen asymmetrisch angebrachten Ballon insofern unterstützt, daß durch Aufblasen des Ballons mit niedrigem Druck (2–3 atm) das sklerotische Material in den

a DVI's Simpson AtheroCath

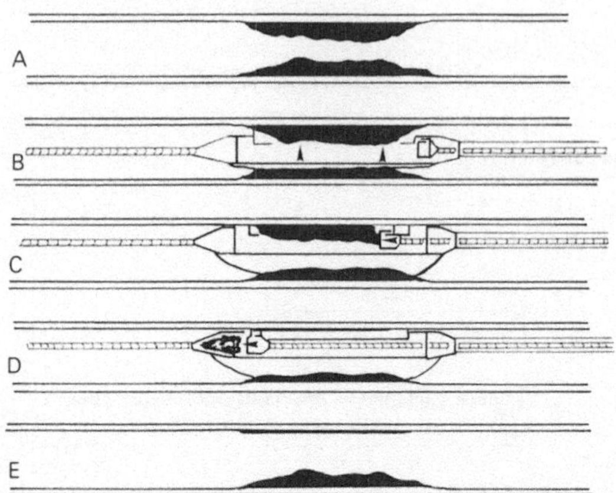

b Theralek's Kensey Catheter

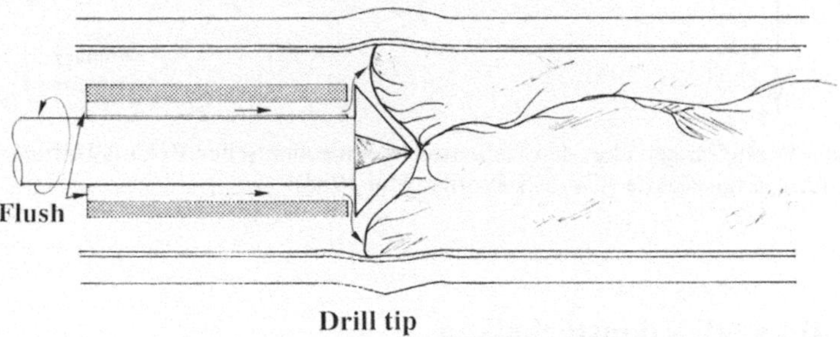

c Biophysics' Rotablator

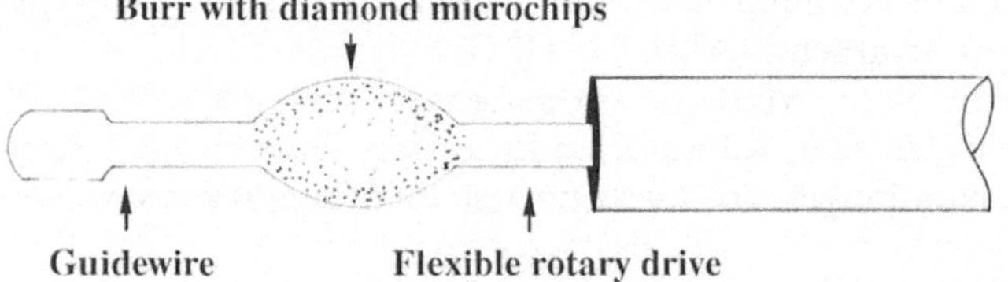

d

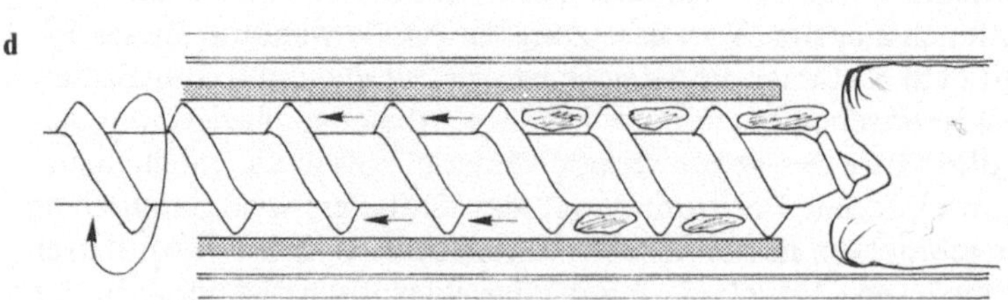

hohlen Metallkolben hineingepreßt wird. Erst dann wird die in den Katheter integrierte rotierende Welle (2000 U/min) eingeschaltet. Durch langsames Vorschieben der mit einer Fräse versehenen rotierenden Welle wird scheibenweise das sklerotische Material in das Hohlende des Metallgehäuses hineingeschoben. Nach mehrfacher Wiederholung des Vorganges wird nach entsprechender angiographischer Kontrolle das System entfernt. Das exzidierte Material kann dann histologisch bzw. elektronisch-mikroskopisch aufgearbeitet werden.

Obwohl diese Technik sehr aufwendig erscheint, sind die klinischen Erfahrungen bisher auch im Koronarbereich recht positiv. Diese Technik ist bei längerstreckigen Verschlüssen allerdings nicht anwendbar. Darüber hinaus scheint die Restenosierungsrate auf der Basis einer Endothelproliferation recht hoch zu sein. Dies gilt insbesondere im Falle einer kompletten Okklusion des Gefäßes mit Werten um fast 50% [13].

Ein vielversprechender Ansatz schien das sogenannte Kensey-Kathetersystem [17] darzustellen. Bei diesem System wird das sklerotische Gewebe durch eine schnell rotierende Fräse (bis 150000 U/min) pulverisiert. Die rotierende Fräse ist in einen Katheter integriert. Zwischen der rotierenden Welle und der Wand des Katheters ist noch so viel Platz, daß eine ständige druckgesteuerte Spülung mit Kochsalzlösung möglich ist, so daß auch die Spitze der Fräse abgekühlt werden kann. Durch dieses Irrigationssystem wird auch das Gefäß im Bereich der Stenose bzw. des Verschlusses etwas gedehnt. Ein wesentlicher Nachteil des Systems besteht darin, daß mit einer hohen Perforationsrate von 10–15% zu rechnen ist. Darüber hinaus ist die Anzahl von peripheren Embolien nicht zu unterschätzen, sie dürfte bei richtiger Einschätzung der Methode ebenfalls im Bereich von 10% liegen [2].

Während die Technik eine FDA-Zulassung für die periphere Anwendung erfahren hat, ist bisher eine Zulassung für die Anwendung im Koronarbereich nicht erfolgt. Daten über die Restenoserate bei Anwendung dieser Methode sind bisher in klarer Form nicht publiziert worden.

Auf einem ähnlichen Prinzip basiert die Technik des von D. Auth entwikkelten Rotablators [11, 24]. Die Spitze des Rotablators ist eine mit feinsten Diamantsplittern besetzte Metallolive, die wie eine hochfrequent rotierende (100–150000 U/min) Minifräse funktioniert (Abb. 2c).

Das System ist erfolgreich nicht nur bei peripheren Obstruktionen, sondern auch bei Koronarstenosen eingesetzt worden. Über gute Erfahrungen mit diesem System hat in letzter Zeit Erbel [5] berichtet. Ein Nachteil an dieser Methode besteht darin, daß bei jedem Eingriff in der Regel zwei oder drei Rotablatorsysteme benötigt werden, da nicht sofort mit einer größeren Olive der Ablationsvorgang begonnen werden kann. Auch mit diesem System ist eine Restenoserate um 30% nicht zu vermeiden. Darüber hinaus muß häufig

◄

Abb. 2a–d. Mechanische Rekanalisationssysteme. **a** Schematische Darstellung des Ablationsvorganges beim Einsatz des Simpson-AtheroCath-Systems. **b** Funktionsmechanismus des turbinengetriebenen Kensey-Katheters. **c** Schematischer Aufbau des Rotablators; **d** Funktionsprinzip des Thrombektomie Rotationskatheters zur Optimierung der Rekanalisation eine anschließende Ballondilatation doch erfolgen. Es bleibt abzuwarten, ob dieses System für eine breitere Anwendung freigegeben werden kann.

Ein weiteres rotierendes mechanisches System für die arterielle Rekanalisation trägt den Namen Transluminal Extraction Catheter (TEC) [6, 22, 26, 27]. An der Spitze dieses Katheters sind zwei Schneideblätter angebracht, wobei die schneidende Fläche von einem Metallgehäuse geschützt wird. Eine batteriebetriebene Motoreinheit bedingt eine relativ langsame Rotation (750 U/min), wobei während des Rotationsvorganges gleichzeitig eine Vakuumaspiration des abgetragenen Materials entlang des Katheters eingeschaltet wird. Dieses System scheint aufgrund des integrierten Führungsdrahtes leichter steuerbar zu sein als das Kensey-System, man muß jedoch mit einer relativ hohen Anzahl von peripheren Embolien rechnen. Darüber hinaus muß man berücksichtigen, daß während des Aspirationsvorganges eine nichtindifferente Menge Blut abgesaugt wird. Weiterhin limitierend ist die Tatsache, daß auch bei dieser Technik eine anschließende Ballondilatation ist fast allen Fällen erforderlich ist. Das Indikationsgebiet mit diesem System bleibt strikt auf die Peripherie beschränkt.

Als weiteres System ist eine Art Spiralbohrer für eine mechanische Thrombolyse [23] zu erwähnen. Der rotierende Bohrer wird leicht gegen das thrombotische Material gepreßt, so daß das abgetragene Material über eine spezielle Gewindeanordnung nach proximal transportiert wird. Dieser Vorgang wird gleichzeitig durch eine Saugeinrichtung unterstützt, die zusätzlich das abzutragende Material an die Spitze des Katheters heranpressen soll. Das relativ langsam rotierende System eignet sich jedoch nur, um frisches thrombotisches Material zu beseitigen. Als Vorteil ist zu erwähnen, daß ein derartiges System prinzipiell auch im venösen Bereich angewandt werden kann oder bei klarer Lokalisation im Falle einer frischen Lungenembolie (Abb. 2d).

Laserrekanalisationstechniken

Allgemeine Betrachtungen

Bei Laserstrahlen (LASER = Light Amplification by Stimulated Emission of Radiation) handelt es sich um Licht, das in Form eines gebündelten Strahls Photonen emittiert.

Das Laserlicht kann im ultravioletten Bereich wie beim Excimer-Laser, im sichtbaren Bereich wie beim Argon-Laser oder im Infrarotbereich wie beim Nd:YAG-Laser erzeugt werden (Abb. 3).

Da es sich jeweils um Licht einer bestimmten konstanten Wellenlänge handelt, ist das Laserlicht immer monochromatisch. Darüber hinaus, im Gegensatz zum natürlichen Licht, bewegt sich das Laserlicht in einem gebündelten Strahl mit einer minimalen Streuung. Da schließlich sich die Lichtwellen in der gleichen Phase fortpflanzen, ist das Laserlicht kohärent.

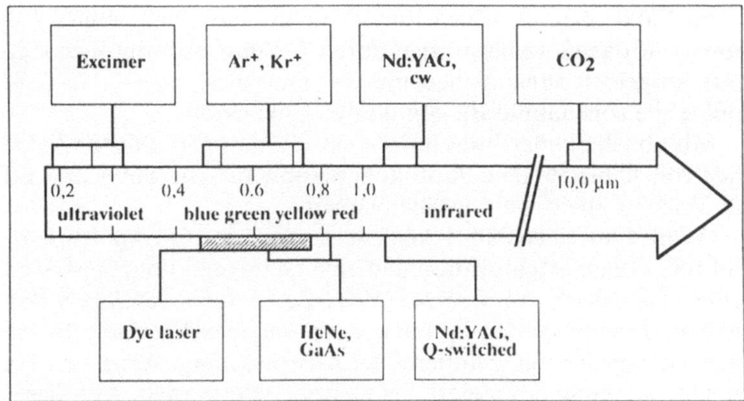

Abb. 3. Die für den medizinischen Einsatz in Frage kommenden Laserquellen

Das Laserlicht wird durch die kontrollierte Abgabe von Lichtenergie bzw. Photonen erzeugt. Dies geschieht durch Anregung von inerten Molekülen bzw. Atomen auf ein bestimmtes, höheres Energieniveau durch Hineinpumpen von externer Energie in das Lasermedium. Die spontan emittierten Photonen können nun selbst in Lasermedium durch induzierte Emission weitere Photonen erzeugen, die dann ihrerseits wieder induzierte Photonen freisetzen. Dieser Vorgang setzt sich lawinenartig fort bis die Photonen das Lasermedium verlassen. Durch zwei Spiegel an beiden Enden eines Resonators kann man die Photonenlawine wieder in das Lasermedium zurückreflektieren und eine weitere Verstärkung erreichen. Bei entsprechender Justierung der Spiegel kann die Photonenlawine bis zu einigen hundert Mal in der Resonator genannten Spiegelanordnung hin und her laufen. Damit ein Teil der Photonen den Resonator des Laserstrahls dosiert verlassen kann, wird einer der Resonatorspiegel teildurchlässig ausgeführt.

Ein Lasereffekt ist an sehr vielen Substanzen, atomaren oder molekularen Flüssigkeiten, Gasen oder in Festkörpern beobachtet worden. Prinzipiell unterscheidet man bei den Laserenergien zwischen sogenannten kontinuierlich emittierenden und gepulsten Lasern.

Die Absorption des Laserlichtes durch das Gewebe hängt einerseits von der Wellenlänge ab, jedoch auch von den charakteristischen Gewebeeigenschaften, d. h. vom Wassergehalt, vom pH-Wert und von der Gefäßversorgung. Es gibt extreme Affinitätsunterschiede in Abhängigkeit von der Wellenlänge des Laserlichtes und dem bestrahlten Gewebe. Das zum Beispiel von einem Argon-System erzeugte Licht wird nahezu vollständig von roten Blutkörperchen, während das CO_2-Laserlicht vorwiegend von Wasser absorbiert wird. Darüber hinaus spielen weitere Faktoren, wie die Energiedichte und die Expositionszeit, eine große Rolle für die Interaktion zwischen dem Laserlicht und dem Gewebe [21].

Im Falle einer niedrigen Energiedichte und einer langen Einwirkzeit des Lasers auf das Gewebe werden durch Lichtabsorption photochemische Prozesse ausgelöst ohne Anheizung des Gewebes. Diese Art von Interaktion könnte die sogenannte Biostimulation induzieren.

Mit abnehmender Interaktionszeit und höherer Energiedichte tritt im Bereich von 40 bis 50° eine Photokoagulation auf. Ein derartiger Effekt wird für das Verschweißen von Gewebe benutzt.

Wenn man einen Nd:YAG-Laser einsetzt, wird im Bereich zwischen 60 und 100 °C eine Denaturation und eine homogene Volumenkoagulation beobachtet. Da jedoch mit dem Nd:YAG-Laser auch wesentlich höhere Energiedichten erreicht werden können, kann das jeweilige Gewebe weit über 100° angeheizt werden mit resultierender Austrocknung. Wenn ein derartiger Punkt erreicht ist, dann vaporisiert das Gewebe sehr schnell. Auf diesem Mechanismus basiert die Beseitigung von thrombotischem, nichtkalzifiziertem sklerotischem Material, gleichgültig, welche Art von thermischen Lasern zum Einsatz kommt.

Im Bereich von sehr hohen Energiedichten sind die elektromagnetischen Felder so groß, daß die Energie ausreicht, um eine Ionisierung der Moleküle zu bewirken. In diesem Energiedichteniveau spricht man von Photoablation und optischem breakdown. Ein derartiger Prozeß kann vom Excimer-Laser bewirkt werden. In einem Übergangsbereich zwischen den thermischen Effekten und der Ionisation gibt es noch eine sogenannte photoakustische Wirkung. Diese Interaktion mag eine Erklärung für manche experimentelle und klinische Ergebnisse mit dem Excimer-Laser liefern.

Bei der Entwicklung eines klinisch einsetzbaren Lasers müssen unabhängig von der gewählten Wellenlänge mehrere Punkte unbedingt berücksichtigt werden. Während die Herstellung von Lasermaschinen in der Zwischenzeit technisch nicht mehr das größere Problem darstellt, gibt es eine Reihe von ungelösten Problemen in bezug auf die Ankopplung der freigesetzten Energie an lichtübertragende Fasern. Darüber hinaus hängt die Qualität der Transmission von den Eigenschaften der lichtleitenden Faser ab. Schließlich ist die Interaktion zwischen der aus der Faser austretenden Energie und dem Plaquematerial problematisch [1].

Damit die Fasern an der Einkoppelseite nicht zerstört werden, muß das Laserlicht wohl fokussiert das Gerät verlassen. Wenn nämlich die Fasern zu nah am Fokus der Linse eingekoppelt werden, mag die austretende Energie so hoch sein, daß die Ablationsschwelle für Quarz erreicht wird und es zu einer Zerstörung der Faser kommt. Wenn hingegen die Fasern zu weit von der geeigneten Position fixiert werden, so kann es zu einer Zerstörung des Cladding kommen und somit zu einer Unbrauchbarkeit des Faserbündels. Diese Probleme sind besonders ausgeprägt für den Excimer-Laser [1].

In bezug auf die Transmission der Laserenergie durch Faser ist zu betonen, daß im allgemeinen die Transmissionseffizienz bei 60–70% liegt, d.h. 30–40% der eingekoppelten Laserenergie wird in Fluoreszenzlicht umgewandelt. Die Transmission scheint hingegen durch den Biegungsradius kaum beeinflußt zu werden, da z.B. für einen Biegungsradius von 2 cm nur ein Verlust von knapp 20% beobachtet wurde [15]. Probleme können schließlich an der Auskopplungsseite auftreten, wenn zu hohe Energien auf das zu ablatierende Gewebe übertragen werden. Zum Beispiel kann eine Zerstörung einer 400 μm Faser bereits bei einer Austrittsenergie von 10 mJ erwartet werden. Experimentell konnte allerdings gezeigt werden, daß die Zerstörungsschwelle erheblich heraufgesetzt wird, wenn Faserspitze und Gewebe von einer Flüssigkeitsschicht getrennt werden [1].

Im Laser-Medizin-Zentrum Berlin wurde mit Unterstützung des BMFT (Projekt Nr. 0706841 9) vor 4 Jahren mit einer systematischen Untersuchung von unterschiedlichen Lasertypen begonnen, die möglicherweise für die Angioplastie geeignet sein könnten.

Es wurden hierbei neben dem Argon-Laser der Nd:YAG-, der CO_2- sowie der Excimer-Laser untersucht.

Bereits die in vitro-Untersuchungen stellten die Anwendbarkeit der thermischen Laser (Argon und Nd:YAG) in Frage. Der CO_2-Laser mußte außer acht gelassen werden, da sein Laserlicht nicht über Faser zur Zeit transmittierbar ist. Wie weiter unten ausgeführt, haben sich in den letzten zwei Jahren unsere Arbeiten weitgehend auf den Excimer-Laser konzentriert, da nach unserer Auffassung diese Laserform die zur Zeit geeignetste für die Anwendung am Menschen zu sein scheint.

Klinische Ergebnisse der Laserangioplastie

Sogenannte thermische Systeme

Die ersten transkutanen peripheren Rekanalisationen mit einem Argon-Laser wurden von Ginsburg [9, 10] vorgenommen. Geschwind [8] demonstrierte, daß auch bei Anwendung des Nd:YAG-Lasers eine Rekanalisation von peripheren Gefäßen erreicht werden kann. In Deutschland berichtete Heintzen [12] über positive Ergebnisse bei der Rekanalisation von peripheren Gefäßen. Diese Autoren benutzten zur Rekanalisation einzelne, sogenannte bare fibers, die in Kathetern integriert waren. Hierbei war eine anschließende Dilatation mit der herkömmlichen PTA-Technik in jedem Fall erforderlich.

Um größere Kanäle zu erzielen und gleichzeitig die Perforationsgefahr herabzusetzen, wurden olivenförmige Metallkappen den Fasern aufgesetzt [14]. In weiteren Entwicklungen wurden statt der Metallklappe Saphirspitzen mit der Laserfaser verbunden [3, 7].

Bei diesen Systemen wird allerdings das Laserlicht nicht mehr direkt auf das Gewebe übertragen, sondern dient nur der Anheizung der endständigen Systeme. Mit dieser technischen Fortentwicklung konnte zwar die Perforationsrate deutlich gesenkt werden, die thermischen Schäden konnten jedoch nicht vermieden werden. Es muß in diesem Zusammenhang kritisch bemerkt werden, daß diese hochtechnologische Form eines Lötkolbens auch wesentlich preiswerter hergestellt werden kann, wenn man als Energielichtquelle nicht ein Laser-, sondern ein Radiofrequenzsystem benutzt.

Unserer Meinung nach liegt der Hauptnachteil aller thermischen Systeme in der Tatsache, daß kalzifizierte Plaques nicht passiert werden können. Gelingt darüber hinaus die Rekanalisation nicht in kürzester Zeit und fällt die Rekanalisationsgeschwindigkeit unter 1 cm/s, so kommt es zu beachtlichen transluminalen Temperaturanstiegen, die Werte um 80 bis 100° erreichen können [1]. Diese Temperaturen sind mit Sicherheit für das umliegende Gewebe schädlich und daher nicht tolerabel. Es muß darüber hinaus berücksichtigt werden, daß Voraussetzung für einen Rekanalisationsvorgang der thermischen Systeme ist, daß das abzutragende Material zuerst karbonisiert und dann vaporisiert wird [3]. Die dadurch entstehenden Karbonisationszonen dürften in vivo beachtliche Probleme verursachen und möglicherweise die hohe Rethrombosierungsrate von bis zu 30 % erklären.

Zusammenfassend kann man kritisch behaupten, daß thermische Systeme nicht weiter für die Rekanalisation von peripheren und noch weniger für die Rekanalisation von Koronargefäßen empfohlen werden können, da bei einer akzeptablen primären Erfolgsrate die Komplikationsrate zu hoch ist und die Anzahl der Restenosen bzw. Reverschlüsse offensichtlich auch nicht akzeptabel ist.

Excimer-Laserangioplastie

Im Gegensatz zu den thermischen continuous-wave Lasern basiert die vom gepulsten Excimer-Laser induzierte Photoablation auf einem sogenannten athermischen Prozeß. Dies hängt damit zusammen, daß einerseits die Penetration des Excimer-Laserlichts in das Gewebe minimal ist und, daß andererseits der lokale Temperaturanstieg, der durch die hochenergetische Explosion bei jedem Puls entsteht, aufgrund der extrem kurzen Pulsdauer in Relation zur relativ langsamen Repetitionsrate von maximal 20 Hz Zeit genug hat, um in das umliegende Gewebe zu defundieren, ohne Schäden anzurichten. Von den vier zur Verfügung stehenden Wellenlängen für Excimer-Laserlicht hat sich die Wellenlänge von 308 nm als die optimale für die Ablation von sklerotischem Material herauskristallisiert [1].

Ein weiterer wesentlicher Vorteil des Excimer-Lasers im Vergleich zu den thermischen Systemen besteht in der linearen Abhängigkeit der relativ hohen Rekanalisationsgeschwindigkeit von der mittleren Energiedichte. Es muß hierbei betont werden, daß die mittlere Energie, die für eine Rekanalisationsgeschwindigkeit von 1,5 mm/s benötigt wird, um das 30- bis 35fache niedriger als bei den thermischen Systemen liegt. Als Konsequenz der niedrigen mittleren Energie, die für die Photoablation benötigt wird, liegt auch die Verletzungszone um den ablatierten Kanal im Bereich von maximal 50 μm, auch wenn man hohe Energiedichten von 75 mJ/mm^2 anwendet [1].

Schließlich konnten in vitro-Untersuchungen an exzidierten Koronararterien zeigen, daß mit dem Excimer-Laser bei einer Energiedichte von 15 mJ/

mm² kalzifiziertes Material abgetragen werden kann. Es ist jedoch zu betonen, daß mit zunehmender Kalzifizerieung die Ablationsrate pro Puls beachtlich abnimmt, so daß bei einem Kalzifizerierungsgrad von 70 bis 90 % die gemessenen Werte nur noch ein Drittel der Ablationsrate von fibrotischem Material betragen [1].

Der Hauptnachteil der Excimer-laserinduzierten Ablation liegt darin, daß die erzielten Kanäle den Durchmesser der benutzten Faser kaum überschreiten. Um dieses wesentliche Problem überwinden zu können, wurden verschiedene Experimente durchgeführt. Es wurde schließlich eine Faser mit einem Außendurchmesser von 360 µm in einen sogenannten open-ended Führungsdraht integriert. Da das Innenlumen des hohlen Führungsdrahtes knapp 1 mm im Durchmesser beträgt, war zwischen der Quarzfaser und der Wand des Hohldrahtes ausreichend viel Platz, um Kochsalzlösung zu injizieren und somit die Spitze der Faser zu reinigen. Mit diesem System wurden Kanäle von immerhin 1 bis 1,2 mm im Durchmesser erzielt, obwohl der Effektivdurchmesser der benutzten Faser lediglich 250 µm betrug (Abb. 4).

Der Grund für die erhebliche Durchmesserzunahme der erzielten Kanäle dürfte darin liegen, daß bei jeder Pulsübertragung über die abtragende Faser das sklerotische Material instantan von einer festen in eine gasförmige Phase versetzt wird. Bei diesem explosionsartigen Vorgang nimmt das Volumen des abgetragenen Materials erheblich zu und es entsteht dadurch eine beachtliche Druckwelle. Man darf nicht außer acht lassen, daß am Ende einer derartigen Faser pro Puls Energiedichten von etwa 1 mW/mm² entstehen. Durch den repetitiven Vorgang (10–20 Hz) entstehen akustische und thermische Wellen unmittelbar vor der Laserfaser. Es kommt offensichtlich zur Entwicklung einer sogenannten Schockwelle, wobei das entstehende Gas versucht, an der Außenseite des Führungsdrahtes auszuweichen. Diese zusätzliche mechanische Einwirkung auf die Materie kann zu einer Disruption des sklerotischen Materials führen als erwünschter Nebeneffekt.

Wie dem auch sei, wäre ein derartiges System nur in Kombination mit einer konventionellen PTCA durchführbar, und das Anwendungsgebiet würde sehr

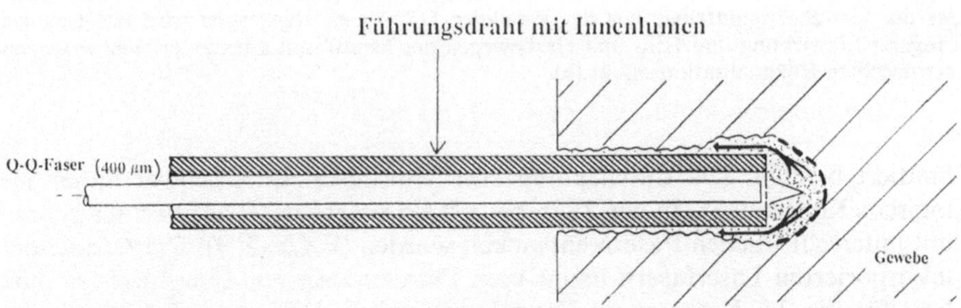

Abb. 4. Schematische Darstellung des sog. Preßhammer-Effektes bei Inkorporierung einer Q/Q-Faser in einen open-ended Führungsdraht. Theoretisch wäre eine Ablation nur im Bereich des kleinen Dreieckes vor der Q/Q-Faser zu erwarten. Die Tatsache, daß man mit diesem System Kanäle bis zu 1,2 mm erreichen kann, basiert auf dem komplexen Vorgang der Photoablation verbunden mit photoakustischen Effekten (Einzelheiten s. Text)

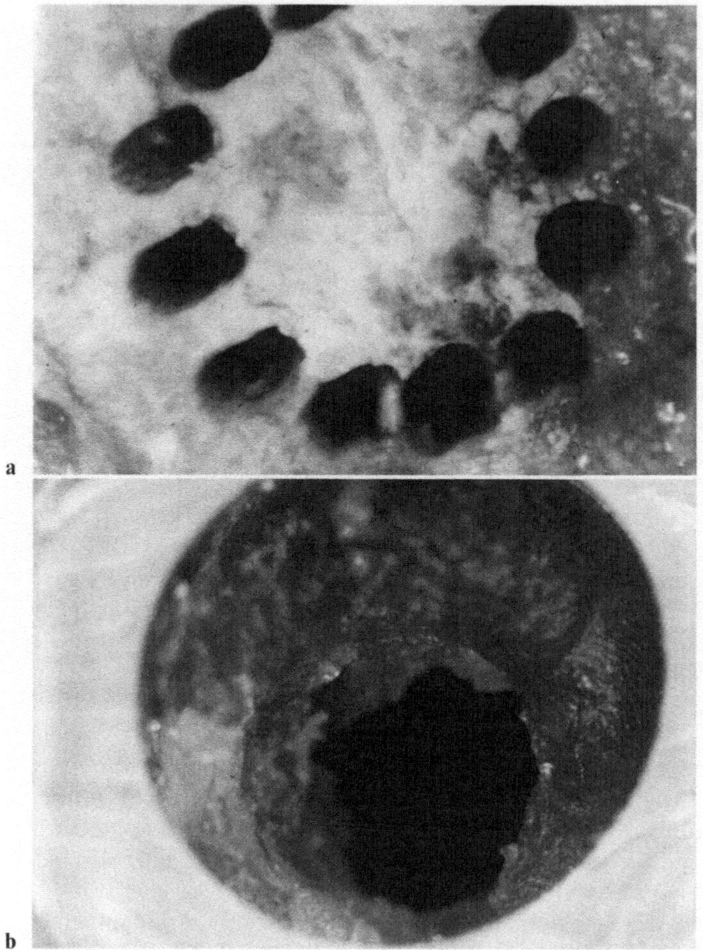

Abb. 5a, b. In in vitro-Experimenten konnte belegt werden, daß mit einem Multifaserkatheter nur Gewebe unmittelbar vor der jeweiligen Q/Q-Faser abgetragen wird (**a**). Erst bei längerer Einwirkung und Hin- und Herbewegen des Multifaserkatheters erreicht man den erwünschten Rekanalisationseffekt (**b**)

limitiert bleiben. Die Entwicklung von Multifaserkathetern war daher die logische Konsequenz. In der Zwischenzeit sind mehrere Typen von Kathetern mit unterschiedlichen Fasern entwickelt worden (F 4,5–F 9). Die Anzahl der inkorporierten Laserfasern hängt vom Durchmesser der Einzelfaser ab und liegt für die 7 F-Katheter im Bereich von 12 und für die 9 F-Katheter im Bereich von 18 kronkorkenartig angeordneten Fasern. Für die Koronarkatheter werden wesentlich dünnere Fasern (Coredurchmesser von 100 μm) benutzt, damit eine höhere Flexibilität erreicht werden kann. Die Anzahl der inkorpierten Fasern liegt im Bereich von 40. Alle Katheter haben einen zentralen Kanal,

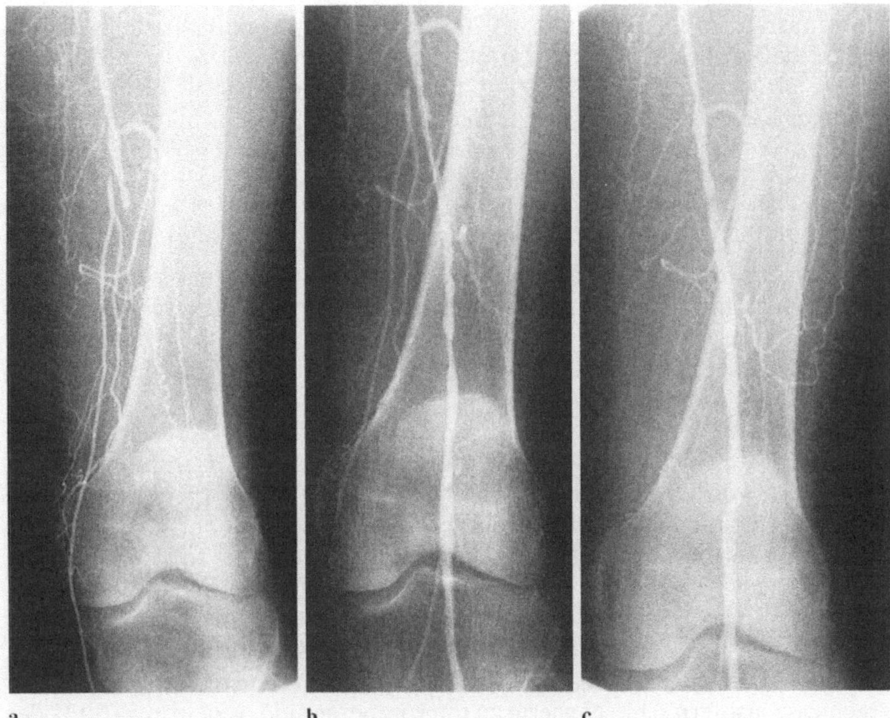

Abb. 6. Kompletter Verschluß der Arteria femoralis superficialis oberhalb des Adduktorenkanals. Vor der Laserrekanalisation war ein Anschlußsegment nicht zu erkennen. Nach den ersten Passagen mit einem Multifaser-Excimer-Laserkatheter (F 7) stellte sich der periphere Anschluß dar. Die anschließende Dilatation diente zur Optimierung des Rekanalisationsergebnisses

so daß sie über einen Führungsdraht bis zur Okklusion problemlos vorgeführt werden können. Der Ablationsmodus eines Multifaserkatheters wird in Abb. 5 verdeutlicht.

Wenn man die experimentellen Untersuchungen mit dem Excimer-Laser zusammenfaßt, so scheint diese Energiequelle für die Ablation von kalzifizierten Obstruktionen deutlich besser geeignet zu sein als die thermischen Systeme. Darüber hinaus reduziert das Fehlen von thermischen Nebeneffekten eindeutig die Gefahr einer Perforation.

Die bisherigen Ergebnisse bei der Rekanalisation von peripheren Gefäßen (Abb. 6) basieren noch auf einer relativ kleinen Anzahl von Patienten [16, 18, 28].

Von Januar bis Dezember 1989 wurden in unserer Klinik 103 Patienten mit der Excimer-Lasertechnik rekanalisiert. Die primäre Erfolgsrate lag im Bereich von 76 %. Es zeigte sich, daß subtotale Verschlüsse ausnahmslos erfolgreich rekanalisiert werden können, während bei totalen Verschlüssen die Rekanalisationsrate im Durchschnitt bei 70 % liegt. Hierbei muß allerdings betont werden, daß während Verschlüsse bis zu 7 cm in etwa 90 % der Fälle, Verschlüsse bis zu 12 cm in 84 % der Fälle rekanalisiert wurden. Liegt die Verschlußlänge im Bereich zwischen 12 und 25 cm, so fällt die primäre Erfolgsrate auf 50 % ab. Besonders schwierig zu rekanalisieren, mit einer primären Erfolgsrate von nur 20 %, sind Verschlüsse, deren Ursprung in unmittelbarer Nähe der Bifurkation lokalisiert ist.

Bisher war es nur in 36 Fällen möglich, ein angiographische Follow-up über 4 bis 6 Monate zu erhalten. Die Okklusionsrate liegt bei dieser Gruppe von Patienten mit 18% in einem sehr günstigen Bereich. Diese sehr erfolgversprechenden Ergebnisse bedürfen allerdings einer näheren und längerfristigen Analyse, bevor man eine definitive Aussage über die Validität der Methode machen kann.

Die Anzahl der Koronareingriffe mit der Excimer-Lasertechnik dürfte mittlerweile weltweit mehr als 600 Patienten umfassen. In Europa wurde die größte Serie von Dr. Karsch aus Tübingen mit 120 Fällen (pers. Mitteilung) vorgelegt. Die initiale optimistische Interpretation der Ergebnisse mußte einer kritischen, eher pessimistischen Einschätzung aufgrund der follow-up-Ergebnisse weichen. Neben einer größeren Anzahl von Vasospasmen während der Intervention, die partiell zu einer temporären Okklusion des betroffenen Gefäßes führten, war auch eine relativ hohe Reokklusionsrate von 30 bis 40 % innerhalb von 6 Monaten zu beobachten. Hierbei muß man betonen, daß die Ergebnisse insofern sehr schwierig zu interpretieren sind, weil in etwa 70 % der Fälle nach der Laserrekanalisation eine anschließende Ballondilatation erforderlich war. Die Anzahl der in den anderen Zentren in Deutschland behandelten Patienten ist noch zu klein, um zusätzliche Aussagen treffen zu können. Sicherlich wird die erforderliche Verbesserung der Kathetertechnik auch zu besseren Ergebnissen führen. Weiterhin ist zu erwarten, daß durch eine verlängerte Pulsdauer von 120 ns auch eine Steigerung der über die Katheter übertragenen Excimer-Laserenergie zu erwarten ist, so daß eine effektivere Ablation die Folge sein dürfte.

Die von den amerikanischen Gruppen erzielten Ergebnisse an mehr als 400 Patienten zeigen eine hohe primäre Erfolgsrate im Bereich von 85 % (Litvack, Los Angeles, pers. Mitteilung). Diese enthusiastischen Berichte müssen allerdings noch kritisch beleuchtet werden, da follow-up-Daten zur Zeit noch nicht zur Verfügung stehen.

Zusammenfassung

Diese kurze Übersicht belegt, daß während der letzten Jahre eine erhebliche Menge Arbeit geleistet worden ist, die das sterile Konzept einer laserinduzierten Ablation von sklerotischem Material in eine realistische klinische Methode umgewandelt hat. Trotz allem möchten wir vor einem verfrühten Optimismus warnen. Wir müssen erst den Beweis erbringen, daß diese neue faszinierende Technologie wirklich einen Fortschritt in der interventionellen Behandlung der

Arteriosklerose bringen kann. Bei allem Optimismus sollte meiner Meinung nach das Ziel dieser Forschungsrichtung nicht sein, Katheter zu entwickeln, mit dem man alle Abzweigungen des Koronarsystems erreichen kann. Der enorme Forschungsaufwand, verbunden mit den erheblichen Investitionen seitens der Industrie, hätte aus ärztlicher Sicht sein Ziel bereits erreicht, wenn die Lasertechnologie einen wesentlichen Fortschritt in der Behandlung der peripheren Durchblutungsstörungen bringen würde.

Ein wesentliches Problem für die wissenschaftliche Evaluierung dieser neuen Technologie besteht darin, daß bevor eine echte klinische Reife erreicht worden ist, ein ständiger kompetitiver Vergleich mit anderen mechanischen Technologien angestrebt wird.

Es ist zu hoffen, daß durch die konsequente Erfassung der von mehreren Gruppen erzielten Ergebnisse man in relativ kurzer Zeit in der Lage sein wird, die Methode zu validieren und deren Stellenwert in der Behandlung von Patienten mit peripheren und koronaren Durchblutungsstörungen besser einschätzen zu können.

Literatur

1. Biamino G (1990) Coronary and peripheral laser angioplasty. In: Interventional Cardiology. Hogrefe & Huber, Göttingen, S 243–260
2. Desbrosses D, Petit H, Torres E, Barrionuevo D, Figueroa A, Heitz A, Grison D, Kieny R (1990) Balloon-assisted rotational angioplasty with Kensey catheter in femoropopliteal occlusions. In: Biamino G et al. (eds) Advances in Laser Medicine IV, Second German Symposium on Laser Angioplasty. ecomed, Berlin (in press)
3. Dörschel K, Biamino G, Brodzinski T, Axel J, Müller G (1988) Comparison of the feasibility of laser angioplasty using heater probes, sapphire tips, and bare fibers. Eur Heart J 9 (Suppl I):331
4. Erbel R, Düber C, Pop T, v Olshausen K, Treese N, Schuster CJ, Meyer J (1988) Ergebnisse der PTCA nach Thrombolyse bei akutem Herzinfarkt. Klin Wochenschr 66 (Suppl XII):119
5. Erbel R, Zotz R, Dietz U, Rupprecht HJ, Stähr P, Meyer J (1990) High frequency rotational angioplasty. In: Biamino G et al. (eds) Advances in Laser Medicine IV, Second German Symposium on Laser Angioplasty. ecomed, Berlin (in press)
6. Feltrin GP, Chiesura-Corona M, Savastano S, Miotto D (1990) Transluminal extraction catheter for the treatment of atheromatous lesions of lower limbs. In: Biamino G et al. (eds) Advances in Laser Medicine IV, Second German Symposium on Laser Angioplasty. ecomed, Berlin (in press)
7. Fourrier JL, Brunetaud JM, Prat A, Marache P, Lablanche JM, Bertrand ME (1987) Percutaneous laser angioplasty with sapphire tip. Lancet I:105
8. Geschwind HJ, Boussignac G, Teisseire B, Benhaiem N, Bittoun R, Laurent D (1984) Conditions for effective Nd:YAG laser angioplasty. Br Heart J 52:484–489
9. Ginsburg R, Kirr DS, Guthaner P, Tolh J, Mitchell RS (1984) Salvage of an ischemic limb by laser angioplasty: description of a new technique. Clin Cardio 7:54–58
10. Ginsburg R, Wexler L, Mitchell R, Profitt D (1985) Percutaneous transluminal laser angioplasty for treatment of peripheral vascular disease: clinical experience with sixteen patients. Radiology 155:619–624
11. Hansen DD, Auth DC, Vracko R, Ritchie JL (1988) Rotational atherectomy in atherosclerotic rabbit iliac arteries. Am Heart J 115:160

12. Heintzen MP, Neubaur T, Klepzig M, Richter EI, Zeitler E, Strauer BE (1988) Clinical experiences in Nd: YAG laser angioplasty in the periphery. In: Biamino G, Müller GJ (eds) Advances in Laser Medicine I, First German Symposium on Laser Angioplasty. ecomed, Berlin, pp 103–113
13. Höfling B, von Pölnitz A, Nerlich A, Bauriedel G, Windstetter U, Berger H (1990) Peripheral directional atherectomy: angiographic and clinical follow-up in 60 patients. In: Biamino G et al. (eds) Advances in Laser Medicine IV, Second German Symposium on Laser Angioplasty. ecomed, Berlin (in press)
14. Hussein H (1986) A novel fiberoptic laser probe for treatment of occlusive vessel disease. Optical Laser Technol Med 605:59–66
15. Kar H, Biamino G (1989) Delivery systems for laser angioplasty particularly considering the safety aspects. In: Safety and Laser Tissue Interaction European Community Medical Laser Concerted Action Programme, First Plenary Workshop. ecomed, Berlin, pp 155–166
16. Katzen B, Schwarten D, Kaplan J, Cutcliff W (1988) Initial experience with an Excimer laser in peripheral lesions (A). Circulation 78 (Suppl II): II-417
17. Kensey KR, Nash JE, Abrahams C, Zarins CK (1987) Recanalization of obstructed arteries with a flexible, rotating tip catheter. Radiology 165:387
18. Litvack F, Grundfest W, Adler L, Hickey A, Segalowitz J, Hestrin L, Goldenberg T, Laudenslager J, Forrester J (1988) Percutaneous Excimer laser angioplasty in humans. Circulation 78 (Suppl II): II-295
19. Meyer J, Schmitz H, Erber R, Kiesslich T, Böcker-Josephs B, Krebs W, Braun PL, Bardos S, Minale C, Messmer BJ, Effert S (1981) Treatment of unstable angina pectoris with percutaneous transluminal coronary angioplasty. Cath Cardiovasc Diagn 7:361
20. Meyer J, Erbel R, Pop T, Rupprecht JH (1987) Derzeitiger Stand der intrakoronaren Ballondilatation. Internist 28:736
21. Müller G, Harnoss M, Kar H, Dörschel K, Berlien H-P (1990) Photoablation a question of wavelength? Laser Technology in Ophthalmology 221:227
22. Newmann GE, Perez JA, McCann RL, Stacks RS (1988) Treatment of peripheral arteries with the transluminal extraction catheter system. Radiology 169 (P): 306
23. Ritchie JL, Hansen DD, Vracho R, Auth DC (1986) Mechanical thrombolysis: a new rotational catheter approach for acute thrombi. Circulation 73:1006–1012
24. Ritchie JL, Hansen DD, Intlekofer MJ, Hall M, Auth CD (1987) Rotational approaches to atherectomy and thrombectomy. Z Kardiol 76:59
25. Simpson JB, Selmon MR, Robertson GC, Cipriano PR, Hayden WG, Johnson DE, Fogarty TJ (1988) Transluminal atherectomy for occlusive peripheral vascular disease. Am J Cardiol 61:96G
26. Stack RS, Quigleyl PJ, Sketch MJH, Stack RK, Walter K, Hofman PU, Philips HR (1989) Treatment of coronary artery disease with the transluminal extraction endarterectomy catheter: initial results of a multicenter study. Circulation 80 (Suppl II): II-583
27. Wholey MH, Jarmolowski CR (1989) New reperfusion devices: the Kensey catheter, the atherolytic reperfusion wire device, and the transluminal extraction catheter. Radiology 172:947–952
28. Wollenek G, Laufer G, Grabenwöger F (1988) Percutaneous transluminal Excimer laser angioplasty in total peripheral artery occlusion in man. Lasers Surg Med 8:464–468

Sachverzeichnis